LIVE SOUND

...cians

PC Publishing

PC Publishing
Export House
130 Vale Road
Kent TN9 1SP
UK

Tel 01732 770893
Fax 01732 770268
email pcp@cix.compulink.co.uk

First published 1996

ISBN 1 870775 44 9

British Library Cataloguing in Publication Data
A catalogue record for this book is available from the British Library

We believe the information and advice in this book to be correct but the author and publisher can accept no liability, consequential or direct, for any use or misuse of the information supplied in this book resulting in damage or loss to any equipment, person or property. The onus is left upon the user to decide for himself how to apply the information supplied in a safe and appropriate manner.

Printed in Great Britain by Bell and Bain, Glasgow

Preface

I've never understood the attitude 'it's only PA it doesn't matter'. Because of the sheer volumes involved it seems to me PA gear should be better, not inferior. We're working at a critical stage, when artists are trying to express themselves live to an audience. The last thing they need is anything that will distract from the performance. An additional problem occurs when people attempt to rectify the poor quality with volume – just making the situation spiral and become worse.

Live sound has to be one of the most challenging areas of the sound industry, perhaps only overshadowed by live broadcast, simply because of the numbers involved. With live sound you only have one chance to get it right. There are no re-takes or pauses. Sometimes there isn't even a rehearsal. You have to mix to perfection a song set you may have never heard before and cope with all the myriad of problems that continually face the live sound engineer – distortion, feedback, getting the monitor mix right for the band, getting the mix right for the audience, stopping the equipment from blowing up, competing with drummers' and guitarists' on-stage volume, and simply just trying to get the thing intelligible and heard.

You're expected to make it sound like a record, even though in the studio they have the advantages of controlled acoustic environments (like space and drum booths), can mix each song ten times, do edits and edit mixes, use grouping and automation, have lots of hands to help, and they can stop and drink coffee as many times as they like. They also may have used different instruments, or at least tuned them, or even used different performers.

As well as coping with long hours, travelling, humping lots of heavy gear, hot sweaty venues, a new set of variables every night, drunken audiences and prima donna artists, you have to get it right or you don't work with that band again.

Then of course if it goes well, you're invisible, the band was just brilliant. If it goes badly, it was because of you – the singer couldn't sing in tune because of your foldback mix, the band couldn't get the feeling because of your mix, etc., etc. It's a thankless job and in the meantime you've got to make sure no one wrecks or steals the gear, or raids the van.

In any case there are lots of back seat drivers who could have done it much better than you, even if they don't know their knob from their elbow. Still there is an amazing sense of achievement and power getting it to happen. It is a real event which will never happen quite like that again, and without you it wouldn't have happened. So sip your coke, down your curry and try and find the next venue on the map. You're a live sound engineer and the show must go on.

This book assumes you're a dedicated sound engineer working for a variety of performers. Of course these days you're as likely to be one (or even all) of the performers as well. In this case apologies for talking about you behind your back! Of course this makes the information in this book even more essential as you'll have little time to experiment or review your own work, so you'll be relying on knowing what will and won't work. Well here are some clues.

Peter Buick

Contents

1

PA versus live sound

Intro

PA has always been regarded as the poor relation of the sound industry and yet it is heard simultaneously by masses of people at high volumes. In this chapter we look at what the requirements of live sound are.

Overview

A PA system in theory is very simple – you just make it louder so that more people can hear it. The traditional PA system as in public address may be just that. Quality isn't really an issue, you just need people to hear the voices intelligibly. But if you think about the quality of most railway station announcements even that would appear impossible.

However live sound is about more than that. It is about providing the quality of a record to the masses – complete with creativity and mood. A mistake with the sound may affect the artistry of the performers. If the vocalist can't hear himself through the foldback monitors, he might sing out of tune, or the band may not play tight because the bass player can't hear the drummer, or the audience can't hear the solo properly because you forgot to turn it up. Then it's your fault – not the performers !

Subjectivity

A major problem with sound is that it is very subjective – it is very much about personal taste and preferences. However, although it is impossible to define the perfect mix, as everyone will have their own ideas, we can usually agree what a bad mix is:

- one that is distorted or noisy, or has important things unintelligible
- one that is muddy in that important instruments cannot be defined
- one that doesn't match the mood of the performance – such as dominant backline on a ballad, or quiet drums on a dance track

but to go further than that is very hard.

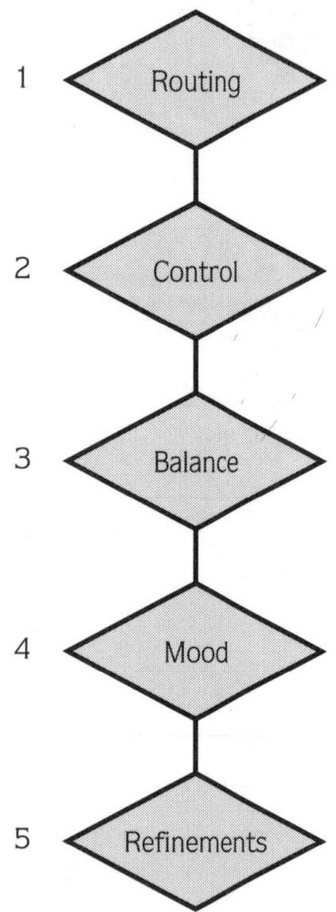

The key elements to live sound

There are many ways to equalise (for instance) a bass, none of them is necessarily better than the others. A sympathetic mix would be one which kept the character that the musician had originally defined. However it may well need adjustment to make a cohesive whole sound as far as the band is concerned. And a nice warm bass sound may be desired, but if it can't be heard as more than a rumble, then what use is it? In such a case, some more slappy, punchy treatment in addition could be advisable. So the art of mixing really has to be separated into five different roles: routing, control, balance, mood, and refinements:

1 Routing – connecting and laying out the gear and getting things to the right places at the right time.
2 Control – making sure that feedback, noise and distortion are avoided and that the levels are sufficient in the monitor mix and for the audience.
3 Balance – making sure everything is audible and is contributing to the sound as a whole.
4 Mood – creating and enhancing the feeling.
5 Refinements – then there is the polish to make things really shine.

Control and creativity

To the majority of people, the extra refinements won't be apparent in the same way as they are to you. It's the difference between a cake and an excellent cake. They're both filling, but one is more satisfying than the other. And most people don't know what makes a particular cake better than another one, they just know what they like. And that's the best way. People who analyse things and comment on a particular aspect rather than the whole, are usually not there to enjoy themselves, they are other musicians and engineers on the lookout for things to plagiarise. Although the opinions of your peers are desirable, it's what the general public thinks that is of more importance. They are the people who pay to be there and will do so again. In short, industry professionals are the worst people to ask an opinion of.

It takes a lot more time and experience and gear to go the extra mile. Sympathetic mixing for each song is a key element in creativity. Or in other words what can you do with the sound to help the performers achieve that mood. This might include a re-balance of levels, adjustment of equalisation or the more obvious use of effects like reverbs and delays.

Then there's the final 10% polish. The bit that makes it better than the record. This might include the use of compressors in a creative rather than a corrective role, spectral mixing to really define every sound, creating false dynamics and riding the faders to get every dB out that you can for a sympathetic mix. This will go largely unnoticed by the audience but sub consciously they will be thinking – hey that sounded good.

If you've got sound coming out of the speakers which isn't distorted or noisy, and everything is essentially equal in level, you're probably 60% there.

Correction and creativity

In essence most elements of PA can be used in two ways: correctively or creatively. For instance equalisation might be used to overcome feedback or to correct the tonal deficiencies in a system. In a creative sense it might be used to change the character of a sound, such as making a vocal deeper. On many occasions these two approaches will clash and a compromise will have to be achieved instead. Between this and the time/live aspect, PA is the loudest compromise on legs !

Planning

We all hate planning, and most people are not naturally organised. However in the high pressure environment of a live performance, it is one of the few ways of coping with the extraordinary, or even the ordinary. If you don't have a system for laying out cables and channels and a rough template for EQ and effects, you stand a good chance of being caught out very easily – especially if something goes wrong.

Troubleshooting

For instance, say the vocal goes down, where does it come up, how can you rectify it quickly and how can you get the same sound quickly? Is it the mic, cable, multicore, mixer channel, effects or sub group ? Is it user error, an equipment fault or just the lead coming unplugged ? Having a system will help you to know what to look for, and if it's out of place you'll know what's up. If you use a different set-up every time you do a gig then how can you tell what's wrong ? You just don't have time to use reason, and trace it through or use substitution techniques for testing – it's all gone quiet !

Working on the fly

In order to work quickly and in stressful situations you are going to need some kind of template system – both for equipment choice and use (i.e. like mic choice and placement) and for mixing techniques (EQ and balance). Hopefully then you'll be able to refine this through the gig to make it excellent. You need to understand what is going on so that you can resolve problems. There is no time for guesswork in live sound, it's got to be educated guess work.

It always amazes me when I see people set up the most bizarre desk layouts. If you lay the desk out left to right as it is on stage, then it makes it a whole lot easier to work out which channel is the stage left vocal mic, or the right kit overhead. It also helps if similar things come up in close proximity. I want all the drums in one place, not split across 24 channels on the board. Yet on many occasions I have seen such disastrous setups.

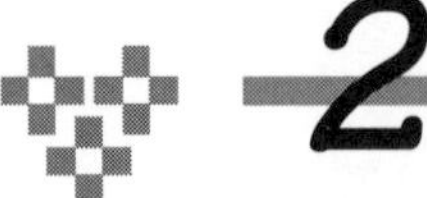

2 The equipment chain

Intro

In this chapter we take a look at the key components that make up a live sound system and how they are integrated, from stage boxes to bi-amplified loudspeaker arrays.

System elements

In any PA system there are a number of elements. The quantity, power, size and cost may change, but the elements do not. To make it simpler, here are the main component areas:

- Inputs – microphones, DI boxes, multicores and stage boxes.
- Control and routing – the mixer which provides gain and level control, equalisation and routing control.
- Processing – outboard effects like compressors, gates, external equalisation, reverbs and delays.
- Amplification – the power amps. This may also include a bi or tri-amplified system where each frequency range is handled separately. This will then include electronic crossovers which split the feed from the mixer into frequency ranges and then feed each amplifier.
- Output – the loudspeakers. People have written books on just this subject. The main factor for us is that there are two sets of speakers – the FOH (front of house) speakers for the audience, and the monitor or foldback speakers for the artists themselves.

Any PA system is just an extension of a greater number of these main elements.

It may also be helpful to consider the physical processes involved in a PA system. Acoustic sounds are first converted into electricity, then they are processed as required, either correctively or creatively, they are then boosted to the required level and are finally converted back into sound by the loudspeaker.

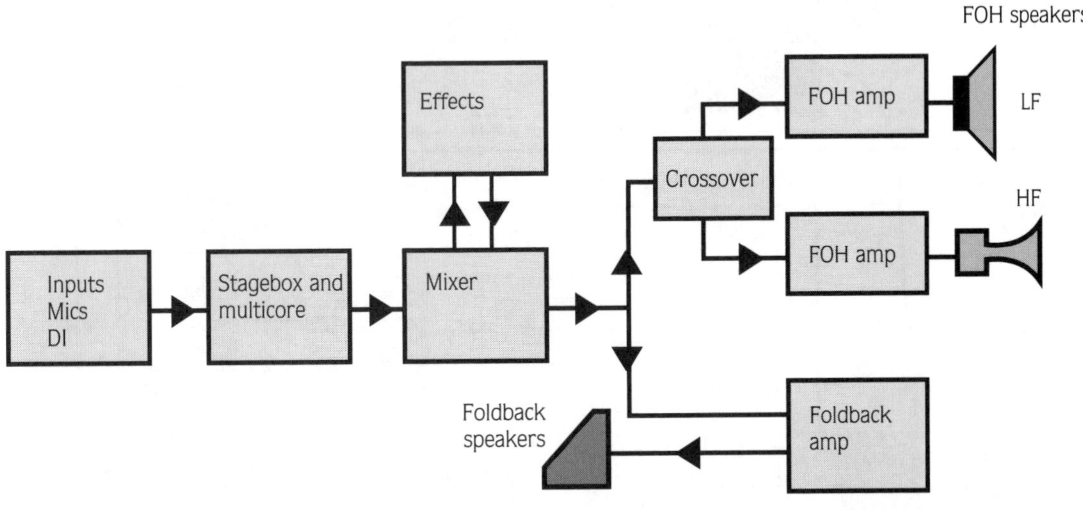

The PA chain

Input
Microphones
Direct injection (DI) boxes
Multicores, stage boxes and snakes

Control and routing
Mixer (front of house)
Crossovers

Processing
Processing and effects equipment
Equalisers
Automation and monitoring equipment

Amplification
Mixer/amplifier (integrated with mixer in smaller rigs)
Amplifiers, FOH and monitor (foldback)

Output
Loudspeakers
Foldback speakers

Functions

To quote the old saying – 'a PA system is only as good as its weakest link'. If you have inferior microphones, it doesn't matter how good the speakers are, the audience just won't hear what they should.

Although it is true that some correction can be made, such as equalising a better response from a microphone, this is always at the expense of other factors – from transient response, noise, gain before feedback, to loss of system efficiency.

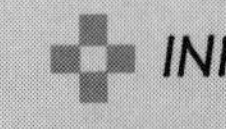

INFO

A 'system' has been defined as a co-ordinated set of components. The microphone is the ear of our system, the loudspeaker is the mouth, and the amplification and processing are the equivalent of the brain.

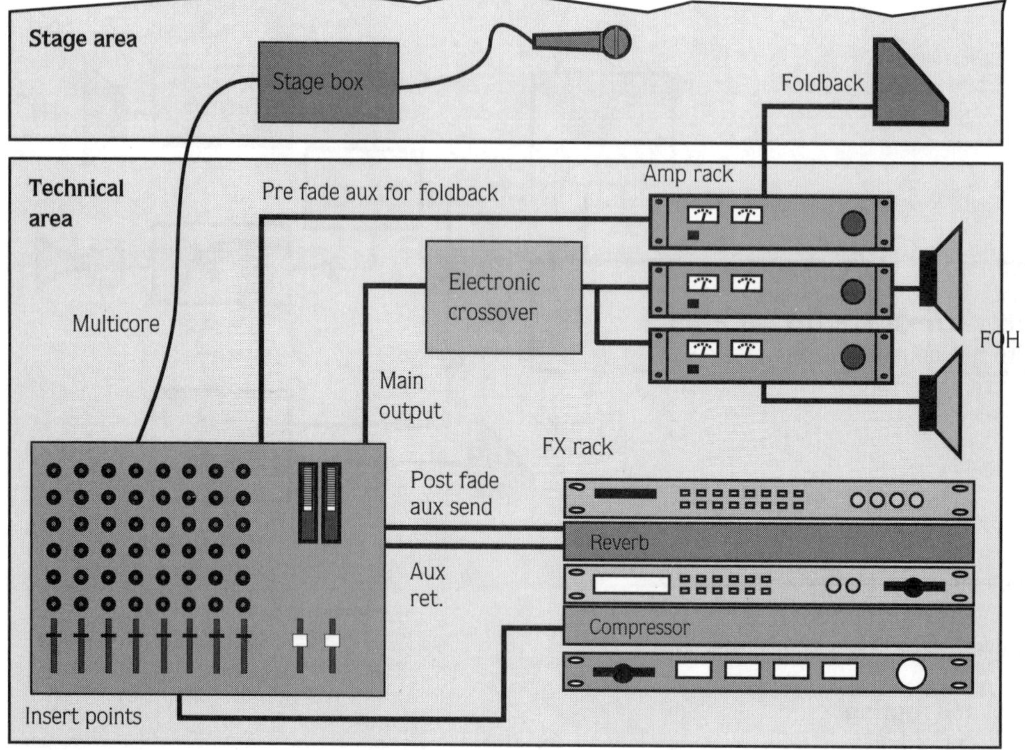

The PA chain – mixer perspective. Notice the boundaries between the stage and technical mixing area (which should ideally be in with the crowd about two thirds of the way back)

Although it might not be possible to get the best of everything straight away, the trade-off between having a reasonable balance to start with, against having to replace inferior equipment later, is a fine one.

See a later chapter for some indications on these financial points.

The PA system in more detail

Let's take a more detailed look at the function of these components.

Microphone

The job of the microphone is to collect the (acoustic) sound and convert it into a form of energy which the equipment can understand – electricity. Because it converts one form of energy into another it is called a transducer. We'll look at how the different types of microphone work later on. But for now that's all we need to know.

Direct injection (DI) box

A direct injection box (DI box) is used to connect line level instruments to the microphone inputs of the mixer. Line level instruments are those which do not need a microphone because they have an electrical output already – such as keyboards, guitars, tape decks and CD players. Record decks are a special case, which also need equalisation (RIAA) as well.

If the line input cables were made available and long enough (i.e. the line inputs were sent down the multicore cable too), then we could plug

There is a financial point at which you get the best value for money. After that the improvement ratio between cost and performance decreases drastically. In other words a product which is twice is expensive probably won't be twice as good – more likely around a 10% improvement will be found.

these sources straight into the line inputs of the mixer, if it had any. However the line inputs on most mixers are not of the low impedance and/or balanced type, meaning that we may pick up interference (radio, hum, etc.) with such a long cable run. Having all microphone inputs available, also gives us the most flexibility in terms of the inputs we can cope with. So for this reason, line inputs are not often available on smaller rigs, and the use of the DI box allows us to convert one for line use.

The job of the DI box is threefold:

1 It matches the impedances and levels of the line level instruments to the inputs of the mixer
2 It removes the potentially harmful phantom power (needed by condenser microphones) from your valuable line level instruments – meaning that they won't blow up.
3 It converts the connectors (a female XLR to a 1/4 inch jack socket)

Attempting to get around any of these factors simply by bodging a lead will probably end in tears !

One DI box is needed for each instrument. A parallel through connector is also provided so that the instrument can be sent to its original source (i.e. guitar to on-stage guitar amplifier) as well.

Special inputs are often provided for high level inputs such as extension speaker outs, so that the amplifier sound can be captured too. Do not plug such high level sources into a DI box unless it has these special high level sockets – or smoke will be arising.

Two types of DI box are available – passive and active. The passive design uses a transformer and needs no extra power, but introduces some attenuation (signal level loss). The active design uses electronic circuits powered from an internal battery. Often, they can use the phantom power supplied from the mixer instead – which saves changing expensive batteries every gig !

In essence a passive DI box is probably a better bet than a cheap active one

It could be argued that the transformer design introduces some undesirable distortion, especially in terms of transient response, but modern designs are pretty good. Electronic versions are more expensive and can introduce electronic hiss themselves. This may or may not be outweighed by the reduction in gain needed from the mixer microphone inputs and the advantages of less interference by sending at a higher level over cable runs.

Multicores, stage boxes and snakes

A multicore is just a flash word for a load of cable housed together in one sheath. To their advantage they house a number of cables in one tidy thick single cable which are identified as a matter of course. A multicore can save hours of cable laying and tracing compared to using single cables. They are unfortunately quite a financial investment and also like most cables they don't take well to being extended later on. Another worry is if one of the cores goes wrong it will be extremely inconvenient. Another good reason to buy more cores than you need.

> **TIP**
>
> A multicore can save hours of cable laying and tracing compared to using single cables

The use of a single big multicore connector on each end is highly recommended for time saving. Although it means you can't cross plug channels easily. If you have a problem, it probably wouldn't matter which channel you can put it on – it's still a problem. Cross plugging only really alleviates an operational viewpoint for the engineer – you're still a channel down. The chances are you'll need to cross plug the microphone end anyway, and this you can do at the stage box anyway.

However if this worries you a lot, you could use a patch panel between the mixer and multicore plug, but remember that you're dealing with balanced low level microphone sources and you'll need to use connectors and cable to suit. The advantage of this is you have the time advantage of a single multicore plug, with the flexibility of using patching only if you need to, if you use a normalised patchbay (see later).

Mixer (front of house)

The mixer is a very important piece of kit. You'll spend 95% of your time bending over it, and the signal may go through it a number of times (considering effects, insert points, etc.), before finally being spat out.

The mixer has two main functions – routing and processing:

1. Routing. From the mixer you can determine which signal goes where. For instance does it go to the foldback, the effects, the stereo mix or some sub groups feeding a special speaker set (or multitrack).
2. Processing. The main types of processing a mixer offers are level and equalisation (tone controls). It also determines how much (if any) external processing (such as reverb) is applied to each source.

Some mixers have built-in effects, but this may not be the best plan. The effects may not be of as high a quality as external options, and will almost certainly be less flexible. The ability to bypass these internal effects and use the effects send with an external unit instead could be appreciated later on, although for the sake of a few sockets and some electronic buffers), manufacturers often seem stupidly ignorant of this.

The Soundtracs Solo PA mixer

In time, as your system becomes more sophisticated, you'll want to develop your own sound (for each artist if you're working with more than one), and having control of external devices will become more important. How pleased do you think the vocalist will be with you, if you put his custom selected mega expensive microphone through the same internal £50 reverb processor as the local Scouts band system ?

Monitor (foldback) mixer

Most people would say that the sound the audience hear is the most important thing, which it is – almost. However, if the artists cannot hear things the way they need to, they will probably not be able to perform at their best. In which case the audience won't really care how good it sounds technically, if artistically it's terrible (i.e. out of tune, out of time, or they miss a synth cue).

It is also largely the monitor mix that the band will be judging you on, and they do pay your wages. Of course the audience sound needs to be good too, but the band are more likely to believe it was good if they have heard something nice themselves.

The main point of the monitor mix is that it is ideally a personal mix for each performer, who often needs more backline and pads (and less frills) than the audience might like. A separate section on the mixer is used to create separate foldback mixes, or in ideal circumstances, a separate mixer and dedicated engineer will control the foldback.

The main problem with foldback is that the speakers are right by the microphones and they are therefore prone to feedback. We'll look at some ways to counteract this later.

The SM 24 twenty four bus monitor mixer

The difference between a graphic equaliser and the equaliser on the channel is that the graphic can adjust several (up to 31 actually) bands simultaneously, while a standard EQ may have only four simultaneous bands (which can often be swept to adopt any one of the graphics sliders and more importantly those values in between).

Processing and effects equipment

Processing equipment can make quite a difference to a system and can certainly be used effectively to create a mood. Nevertheless, it is very hard to compensate for deficiencies in the system, although it is easier to take things away.

Besides obvious effects like flanging, chorus and delay, a bit of reverb can help to give that studio quality feel, even though the venue may be live enough already. More importantly perhaps, it will give the performers more confidence.

Compression and gating effects can really help to clean up the sound and give it a real edge. We'll look at these in detail later.

Equalisers

An equaliser can be used in two ways – artistically and technically. Artistically it can create an effect or mood; technically it can compensate for the acoustics of the venue and audience, or maximise on gain before feedback.

Here we are mainly talking about graphic equalisers, so called because they have a slider control (fader) for each frequency that they can adjust, and the physical position of the sliders graphically shows what it is doing to the sound. Despite the glossy adverts, the proper setting for a graphic is not necessarily a nice wave or U shape.

Graphics are useful tools but can seriously alter the sound and even make it worse. So care and discretion need to be used with this device. If you're having to adjust the sliders widely, there's probably another problem somewhere you should be dealing with.

If you find you're using similar EQ settings on all the channels then it can be a lot more efficient and less noisy to transfer this setting to the graphic and add it globally.

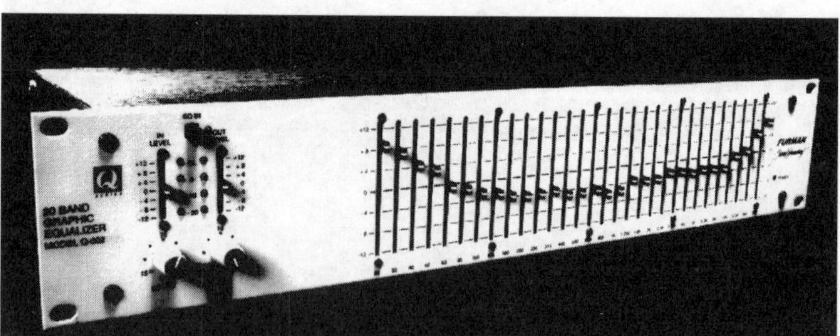

The Furman 30-band graphic equaliser

Amplifiers

The job of an amplifier is a simple but important one – to make things louder. However this is never a perfect process, and all sorts of factors come into play – e.g. which speakers they like talking to to start with.

The technical term for this is the damping factor but what it means is that certain amps work better with certain speakers. When the amplifier sends a current through the speaker's coil, it creates an electromagnetic field that is alternately repelled and attracted to the permanent magnet that the coil sits in. This makes the coil, with diaphragm attached, move – and the moving diaphragm produces sound. But when the coil moves, it also cuts the magnet field of the magnet and hence produces an electric current of its own, mainly in opposition to the amplifier. It is quite a small current in comparison, but the amplifier has to deal with it. One rule of audio states that 'things behave differently at different frequencies'; the speaker has its own impedance to following and generating electric currents.

In short this means that some amplifier/speaker combinations sound terrible in comparison to others. There's nothing wrong with either part, just the way they interact.

Crossovers

Although you might be a great footballer and can write a nice neat letter to your granny, chances are you'd find it hard to do both at once. Equipment has the same concentration problems too – especially speakers.

By designing speakers to work within more limited frequency ranges: i.e. woofers, mid-range units and tweeters, and then separating the amplifier's signal into different corresponding frequency bands, we can produce a more efficient sound system. Hence we have low frequency signals sent to the woofer, mid-frequency signals sent to the mid-range speaker, and high-frequency signals sent to the tweeters.

Any integrated speaker cabinet will contain a passive crossover network (even if it's just a resistor and the natural impedance of the coil and mechanics), but a better sound can be achieved with external means.

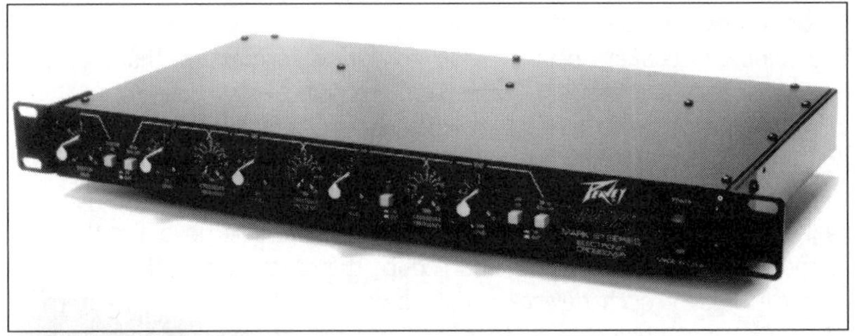

The Peavey V4X electronic crossover

Bi-amping

Bi-amplification is such a process where the sound is split into ranges of bass and mid-range/treble. Each half is given its own amplifier (so that it can concentrate too), so that maximum efficiency is achieved. It takes a lot less power to achieve the same result with treble compared to bass. As bass needs to move such a lot of air it also needs more power, in addition to the fact that human hearing is more sensitive in the mid to treble range.

As well as leaving each component to concentrate on what it does best, it means we can optimise the amplifier power for each frequency range (the treble needs a lot less power, around ten times less in fact). As well as this, it means that each range won't be subject to the inadequacies of the other. So for example, if the bass amplifier is overdriven and starts clipping, that HF distortion won't be fed into the tweeter. Also if compressors are used after the crossover to limit the power, they will behave as if naturally frequency masked in their operation.

Loudspeakers

The loudspeaker couldn't be a much more basic piece of technology – it comprises mainly a coil of wire suspended in a magnet and glued to a piece of paper – yet the science of speakers is enormous and it appears to be largely empirical (suck it and see).

The job of the speaker is to convert the current coming from the amplifier back into sound, by vibrating the diaphragm in sympathy with the electric signal. There are a lot of odds against it doing so – including its own size, weight, rigidity, mounting, air resistance from free air and cabinet design (it's also affected by temperature), and back EMF (damping factor) by moving in its own magnetic field,.

There are some basic rules of physics that apply to speakers. One is that its size, and the volume of the cabinet it is housed in, basically dictate how low a bass frequency it can reproduce. Then the shape and paths inside and the inclusion of any ports and vents all play their part too.

Speakers also have their own character which reveals itself in frequency response and dispersion. In other words it probably won't be as responsive to all frequencies, with variations up to 10dB being considered good.

Foldback speakers

The foldback speakers deliver a mix to the performers so that they can hear what they and the rest of the band are doing.

As foldback speakers are close working, they don't need to be tremendously powerful, but care needs to be exercised with possible feedback.

A recent development in foldback has been the use of in-ear devices. As well as providing less clutter on the stage and sufficient level for the performer, they negate feedback and reduce the on-stage muddle of sound. As the prices of such devices come down, maybe we'll all benefit from them. Do remember though that they can still deliver sufficient level to damage hearing and you have less of an indication of it.

Gadgetry

As PA technology has become more sophisticated, all sorts of gadgetry have been developed. Devices which can analyse the status of your amplifiers and tell you if the output is distorted or not, and devices that can analyse and correct feedback automatically, are all now available.

Foldback monitors (wedges)

Remote control

Remote VCA control of amplifiers has become more commonplace as it reduces the clutter and often allows shorter cable runs between the amplifier and speaker, which, with the levels involved here, can be quite a benefit. Remote VCA control means that the amplifier can still be adjusted easily as necessary.

With the addition of remote monitoring, it makes remote amplifier placement a reality rather than a nightmare. If the amplifiers are distant you cannot set them up properly for the venue, or see if they have gone down. Hot-swapping of amplifiers to rectify such failures can still be a problem with remote location, but such a system is usually based on multiple amplifiers, and the loss of one amplifier would not be such an issue. Remote patchbays, spare sets of gear, or simply legging it through the audience to the amp rack, can be employed otherwise.

Spectrum analyser

Graphical monitoring devices called spectrum analysers are available fairly cheaply now, and these can help to indicate visually what's going on sonically. Whilst these devices are very useful as tools, they should not be relied upon completely. Although they can help you to achieve a flat-response, low-feedback EQ setting, it may sound horrible and still ring. Your ears are the best judge of most things audio.

Automation

Automation isn't a natural word to use in the context of a live event. The whole concept of 'live' is that it is a unique spur of the moment event not to be repeated. Truth of the matter is that it is more likely to be an attempt to recreate something rehearsed, in front of lots of people. In this scenario, anything which can help realise this is welcome.

By providing any system of automation, routine and often complex operations can be performed with security, leaving the more artistic and enjoyable parts of the job for the live operator. Automation of course can be introduced at any stage, aka 'on the fly', so a live response can still be handled. For instance say you wanted a quadraphonic panning effect at a point in the song. This can still be flown in manually, but the control of the pan can be pre-recorded.

Similarly with level, EQ, or effects changes between numbers; these can be pre-programmed for faultless instant recall. In other words, automation systems can act like an extension to the engineer's memory and hands. He still controls their programming and recall, but, once done, the results will be consistent and repeatable.

3

Systems and applications

Intro

The specifications of a system often need to change with the application. We consider some typical applications and systems, including live sound for small bands, fixed and mobile installations, expandability, system performance and efficiency and how many watts you actually need.

The first question about a PA system is its application. If it is just to reinforce a business presentation to a large audience, then its main purpose is to spread the sound so that everyone can hear it at the same sort of volume (around 85 dB) – even those at the back. In this case it would be inappropriate to overpower the audience with power.

Cabaret music

For a cabaret type event where people may wish to carry out other activities such as talking, an average level is required, perhaps around 95 dB.

Small groups

For small groups, such as acoustic bands or singers, the PA may be required to deliver a slightly more impressive volume than acoustic singing, perhaps around 110 dB.

Rock music

For rock music the volume needs to be absorbing and powerful, so much higher levels are needed, often reaching the threshold of pain at 130 dB. Legally of course, the volume should not exceed 95 dB as is set by many councils with their venue cut out globes.

Classical music

For classical music, dynamic range is quite an issue, and we need to be able to hear a wide range of sound levels, from a solo violin to a full crescendo including timpanis. Around 120 dB on peaks would be expected here, with a strong bias on natural sound and low ambient noise.

Dance music

For dance music, a similar idea to rock music is required – arguably slightly less but more localised and with an accent on feeling the bass.

Especially when you consider that people may be in there for many hours longer than a gig and will usually be under the influence of something. They may also still wish to talk, or at least shout, at each other. Again legally 95 dB would be a target, but in practice more likely 118 dB.

These levels are obviously only guides and will be adjusted to taste. Things to bear in mind, other than the legalities of potentially deafening people (which brings us back to 95 dB), are that level is a very comparative thing. The ear soon adjusts itself to pay attention whatever the level. It may be very loud when you first walk in off the street, but you'll soon adjust and forget the impact of the volume until you go out again.

What is impactual then is contrast. If something is quiet and then gets loud, the change is perceived as a greater event than the mere volume. A solo singer in an acoustic venue can still deliver quite an impact by going very quiet, to make his loudest singing more of a change.

This perhaps is the most important thing about live sound. Unfortunately it isn't very easy to convince anybody about it. Ask your average guitarist or drummer to consider dynamics and he thinks you're talking about cars. He feels impressed at the larger than life noise which makes pretty much anything sound good, or at least impressive – for a while anyway.

If we could put more dynamic contrast back in to live performance, the sound industry would have come out of the stone age. Till then, you might have to turn it up.

As an experiment you could try turning the main volume down through a gig and you'd probably find no one really noticed. The only problem is competing with the things you can't turn down, like drums and guitarists with backline. The sooner guitar amplifier manufacturers start fitting MIDI controlled (remote) VCA facilities the better for the live sound engineer.

PA system problems

Regardless of what components we use to deliver the PA, the basic questions are still the same.

The PA system test

There are several criteria:

1. Is it loud enough without distortion to compete with the backline and deliver the level/impact required?
2. Can it compete with ambient noise and supply the required dynamic range?
3. Can it cover the whole audience and still deliver intelligibility (reverberant fields)?
4. Can we get this level without feedback when using microphones?

The technical issues which deliver these criteria (or not) are power, efficiency, dispersion and separation.

Power, efficiency, dispersion and separation

Choosing the right PA system is an art in itself. PA is largely about compromise and it's a question of drawing the line at what is acceptable. For instance to get a good bass response requires large speakers, and louder speakers tend to have larger and heavier magnets in them. But most people find a smaller lightweight rig more convenient. People prefer to have less gear and connections and often opt for a combination mixer/amplifier, but much greater control can be had from separate components. For instance can the combo system be expanded with better effects, does it have inserts for external processors, can it cope with bi-amplification and can you cross patch the amplifiers if the FOH section goes down ?

Power

A basic guideline for the power required is 1 watt per person in the audience. But this assumes that you're not trying to compete with loud backline and drummers who thrash the kit like their life depended on it. If they won't turn down for the smaller venue, then you have little choice if you want the vocals to be heard.

A kit can easily put out 120 dB at 1 metre (more likely 142 dB actually), so you can see what you're up against. With an average speaker sensitivity of 96 dB at 1W at 1metre, it would take a 256 W power amplifier to compete with it.

You could of course try moving the drummer back 256 feet (a 3 dB drop for every doubling in distance) to reduce his level to a more modest 96 dB (although it will sound very reverberant and distant of course), or just ask him to play quieter. It's more practical just to turn up the PA though.

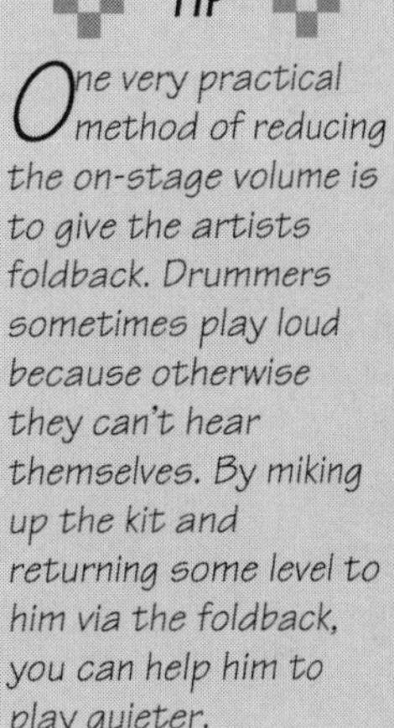

Efficiency

A main problem with defining power for a PA is its efficiency. To understand this we need to first understand the concept of the dB. Look at the dB section for further details of this.

In practical terms the following example may help. Say we had a choice between a 400 W speaker and 200 W speaker both costing the same. You'd go for the 400 right ? Well now let's look at the specification closer. The efficiencies of the speakers are:

200 W: 106 dB for 1 W at 1 metre (at 1 kHz)
400 W: 96 dB for 1 W at 1 metre (at 1 kHz)

What this means is that to get exactly the same volume from the 400 W speaker we would need to feed an extra 10 dB of power in to it from a higher powered amplifier. To get a 10 dB increase in power is a ratio of ten. That means we would have to use ten times the power of the amplifier to get the *same* result – i.e. 100 W to 10 W amplifier. I hope you'd now reconsider your choice of speaker.

There are of course other factors involved, such as do you like the sound of it, its frequency response and the efficiency at other frequencies. So you still need to listen to both systems to choose.

A 3 dB improvement means a ratio change of two, so every time you can get an extra 3 dB from something, it's worth going for. To recreate this with the power amplifier will mean a much bigger amplifier – twice as big in fact. This applies to microphones, speakers and electronic devices before distortion sets in.

The decibel (dB)

The decibel is a measurement of a ratio. The idea is that a known and fixed reference is used which makes it easy to compare things. The other advantage is that dBs can be simply added and subtracted rather than having to work out complex mathematic operations.

Voltage dB

The voltage ratio is usually to a reference of 0.775 V. Some Japanese equipment uses the dBV which is referenced to 1.0 volt, but this is not common in PA circles.
The formula used to work out voltage dBs is:

dB = 20 log v / Vref

This means that:

If the mixer puts out 0.775 V, then the level is 0 dB.
If the mixer puts out 1.55 V then that equates to +6 dB.
If the mixer put out 7.75 V (say on peaks) that is 20 dB.

If we wanted to work out the other way round, i.e. how many volts so many dBs was, then formula transposition shows us that;

if dB = 20 log (v /Vref) then
dB /20 = log (v/Vref)
antilog (dB/20) = v/Vref
v = antilog (dB/20) x Vref

So	+4 dB	= 1.23 V	pro line level
	+3 dB	= 1.095 V	
	+6 dB	= 1.55 V	twice the 0.775 V voltage
	+20 dB	= 7.75 V	ten times the 0.775 V voltage
	–10 dB	= 0.245 = 245 mV	
	–7.8 dB	= 0.316 = 316 mV	semi pro line level
	–20 dB	= 0.775 = 77.5 mV	guitar level
	–50 dB	= 0.00245 = 2.45 mV	mic level

(Note that –10 on semi pro Japanese gear is referenced to dBV so is actually 316 mV or –7.8 dB)

This formula is also used for sound pressure (SPL) with a reference of 0.0002 dyne/cm^2 (or 20 μPa)

SPL loss with distance can also be worked out using this formula

i.e. 96 db at 1 W at 1m – SPL at 10 m is

20 log (10/1) = 20 dB

Therefore SPL at 10 m is 96 dB – 20 dB = 76 dB

Power dB

The formula to work out power dB ratios is:

dB = 10 log (W / Wref)

where the reference is 1 mW into 600 ohms. So

10dB = ratio of 10
6 dB = ratio of 4 (approx.)
3 dB = ratio of twice (approx.)
–3 dB = ratio of a half
–6 dB = ratio of a quarter
–30 dB = ratio of 1/1000 (x 10^{-3})

Sound power also uses this formulae with the reference of 10^{-12} watts.

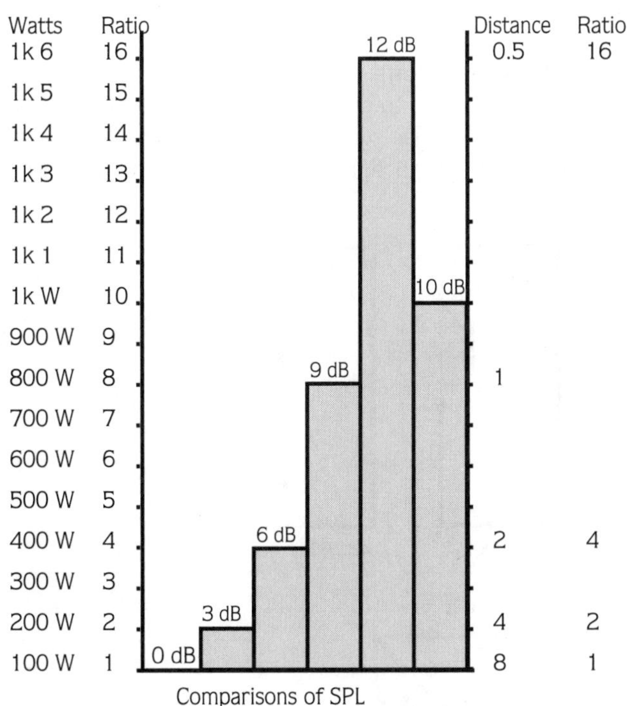

Graphical representation of dB power ratios.

Dispersion

The main factor controlling the number and type of speakers required is down to dispersion and coverage area. Dispersion is how well the sound can cover a certain area and distance.

With a wide dispersion speaker we can run in to problems of uneven frequency spread and also poor intelligibility because the sound is reflected off surfaces and increases the reverberant field. With a narrow dispersion speaker we can control reflections more, but we will probably need more speakers to cover the whole area. Then we run into problems of interference between the speakers causing all sort of frequency response problems. As I said earlier, PA is all about acceptable compromises.

So we have a concept of dispersion, a rough idea of the power we need and we know we'd like efficient speakers, but what type?

There are several types of enclosure design. They each have their own merits and drawbacks and largely we must be led by the fact that the manufacturer's choice is correct. However let's see what the options are.

Dipole

The simplest type of enclosure is no enclosure. By mounting a speaker on an open backed baffle, we create a dipole speaker. These are quite common in guitar combos and for in car entertainment. The dipole is very easy to construct. The main problems are interference between the front and back of the unit, which tends to reduce the bass response. This system tends to be only 5 – 10% efficient at converting electrical power in to sound.

Sealed (acoustic suspension)

The next logical enclosure is to completely seal the rear of the speaker to stop this interference. We get excellent bass response but it is a very inefficient cabinet as it needs to fight with the air in the enclosure all the time.

Various types of loudspeaker enclosure

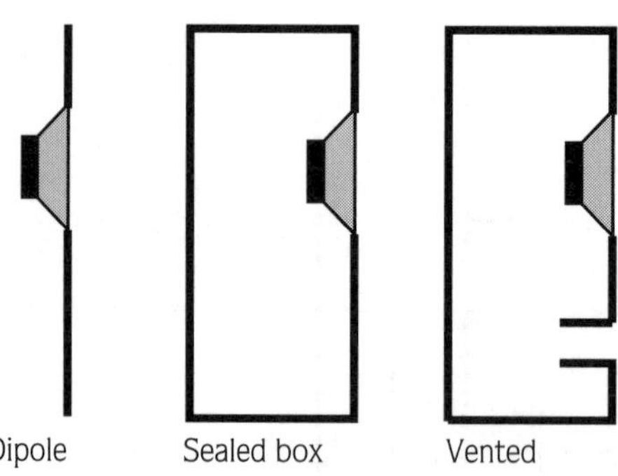

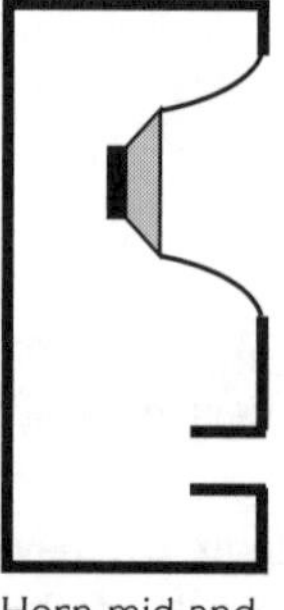

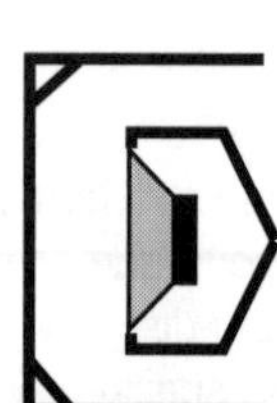

A typical vented speaker and horn

Vented (bass reflex)

The vented speaker is a sealed speaker with a special vent which allows the air to flow but controls any interference. The results from vented designs have been very much a hit and miss affair. Not until the 1970's had the 'science' of vented design been formulated – mainly by two people Nevile (AN) Thiele and Richard Small. We will look more at their work in a later chapter. Needless to say that the design of the enclosure and vent is critical in the proper working of the speaker.

By using a weighted cone in the driver, a so called passive radiator design can be achieved as a sub type of the vented design.

Horns

Ever since the phonograph and megaphone were invented, people have understood the use of a horn to direct the sound. By focusing the energy of the sound, an apparent amplification can be achieved with a horn.

This horn concept can be used in all ranges of the spectrum, and horn bass and mid drivers are almost as common as the HF (high frequency) variety. In bass cabinets, the horn has to be so long due to the wavelengths of the sound involved, that the horn is often folded inside the cabinet, to make the enclosure size manageable. A horn can offer a 15-30% efficient design.

The reason that a horn design works is that it focuses the energy in a smaller directional route. So instead of the energy having to dissipate over an increasing square area, the sound can be concentrated in the direction of the horn.

Combination – horn mid and vented bass

Naturally combinations of the above designs have evolved. The horn mid with a vented bass is one such adoption, ideally suited for general and vocal use where good projection is required. The design of such units is very complex and tends to lead to a rise in response at the crossover frequency between the horn and vent elements.

Combination – horn bass and sealed mid

Using a horn bass and a sealed mid offers the alternative of strong bass emphasis, ideally suited for bass instruments, drums and organs. It requires a very complex design and is slightly inefficient compared to the horn midrange design, but works better for bass.

The following table summarises the benefits of each enclosure design:

Enclosure characteristics table

System type	*Figure of merit (K)*	*Design complexity*	*Response*	*Characteristic*	*Usage*
Dipole	greater than 1	very simple	rolled off bass	projects well	guitar combos
Sealed box	1.5 – 2	simple	flat or rolled off bass	tight punchy	guitar, midrange and vocal
Vented box	3 – 4	complex	flat	smooth hifi with solid bass	floor monitor and vocal
Horn	1 – 2	very complex	flat or bass rolloff	tight punchy mid bass	LF building block
Combo (horn mid)	3 – 4	complex	response rise at horn range	good projection	general use and vocal
Combo (horn bass)	1 – 2	very complex	elevated bass	strong bass	bass instruments

So there we have an overview of the types of enclosure, so what do we do with them ?

Split stack

The most common layout for speakers is a split stack. A set of speakers stage left and stage right. The bass bins are at the bottom because bass is pretty omni directional, then the mids and horns which are more at audience ear height, due to their directionality.

The problem with this system normally is that to get a healthy level at the rear of the audience, it has to be very loud at the front. This is because sound decays at a rate of 3 dB for every doubling of distance, so although it might be 100 dB at 1 m (3 feet), 32 m (96 feet) back it will only be 85 dB.

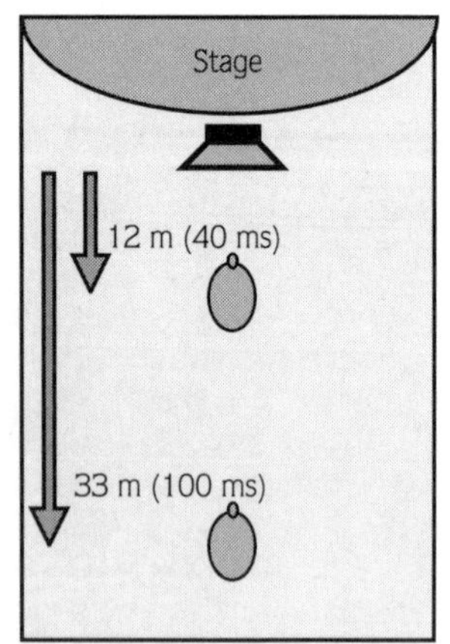

Sound takes time to travel so it reaches the back of the hall later. Delay compensation may be required

To equate this to the real world, 100 dB is the level of loud classical music or a guitar amp at 1 m, to 85 dB for an acoustic guitar or inside a sports car at 55 mph.

Also, as high frequency sound has low energy content it will tend to get absorbed on the way. Another problem is the balance between the direct sound from the speaker (the direct field) and the sound coming from reflective surfaces like walls (the reverberant field). The reverberant field is comprised of many complex time delayed and frequency coloured reflections which when combined with the original sound, often produce an unintelligible muddy sound. The only way around this is to reduce the amount of it.

This can be done with multiple speakers nearer each section of the audience, and in very large venues, speakers along the hall will be used. As sound travels at around 30 cm (1 foot) per millisecond, these speakers may also need to be time delayed to reduce the echo effect between the local and distant speakers, or else again the sound will suffer. So for a distance of 12 m (40 feet), a 40 ms delay can result near the ADT and

Speakers are flown at this height so that they can be directed towards different parts of the audience

chorus region. At 33 m (100 feet) a 100 ms delay occurs which is almost a distinct echo and will certainly appear as a reverberant effect. The local speakers need to be delayed slightly more than the 1ms per 30 cm, to maintain the direction perspective of the event.

Alternatively, so called long throw speakers can be used. These have a narrow dispersion but a much longer range and hence can be aimed at the rear of the hall directly where needed and largely eliminate any surface reflections. The problem with this arrangement is that more speaker units may be needed to cover the area and, if care is not taken, these will interfere with each other.

This is the concept behind the central cluster design. It consists of a flown collection of long throw speakers, each aimed to cover an area. Sub woofers can also be employed at ground level to save suspension weight, as bass is largely omni directional. As all the speakers come from one place, interference between units can be optimised.

TIP

So with reverberant (live acoustic) venues, we need speakers with a narrow dispersion so that we can aim a larger quantity of them where required. Care needs to be taken to avoid interference effects between the speakers. For less reverberant areas, fewer wider dispersion speakers can be utilised, saving on bulk and speaker nterference.

Separation (bi-amplification)

Another factor in a PA system is how well each unit can handle itself. We've seen how speakers and sound can mechanically interfere with each other, and the same is true of most PA system elements in electronic terms too.

Firstly, we don't want to feed speakers with frequencies we don't want them to handle. This will only reduce their efficiency as they try (and often fail) to reproduce the sound. For instance feeding HF to a bass driver (woofer) will have little effect, while feeding an HF unit with bass will tend to blow it up.

In speaker cabinets there is usually a passive crossover whose job is to separate the frequency components to the appropriate drivers. It does this by using a passive (needs no power) network of capacitor and inductor electronic components called a crossover. In very cheap systems, resistors may be used to reduce the power to drivers, but this wastes the power from the amplifier and turns it in to heat. Passive crossovers are built in to the speaker cabinets and are placed between the power amplifier and the speaker drivers themselves.

The passive crossover is very cheap to build and does its job, but better results can be achieved by using active electronic versions of them. This then allows the output of each band to be fed to its own dedicated amplifier which can be optimised for its job. This means that as the amp is only fed with the signal range it is interested in, any clipping occurring in other regions will not affect it, as would be the case with global clipping. Also it means the power and transient response of the amplifier can be used to best effect, rather than it having to deliver power to just the loudest global signal.

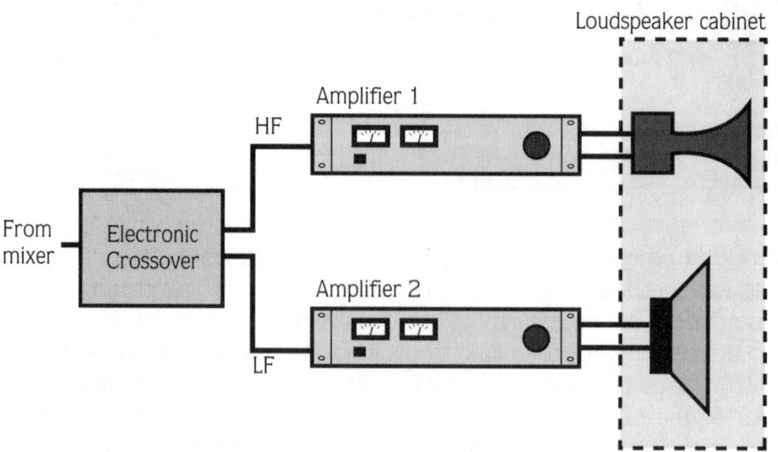

A biamping system. An electronic crossover can split the spectrum to feed dedicated range amplifiers producing a more controlled sound. Two and three way (bands) are common. Note how the electronic crosssover is placed between the mixer and the two amplifiers

Tri-amplification

Similarly splitting the signal into three sections is called tri-amplification.

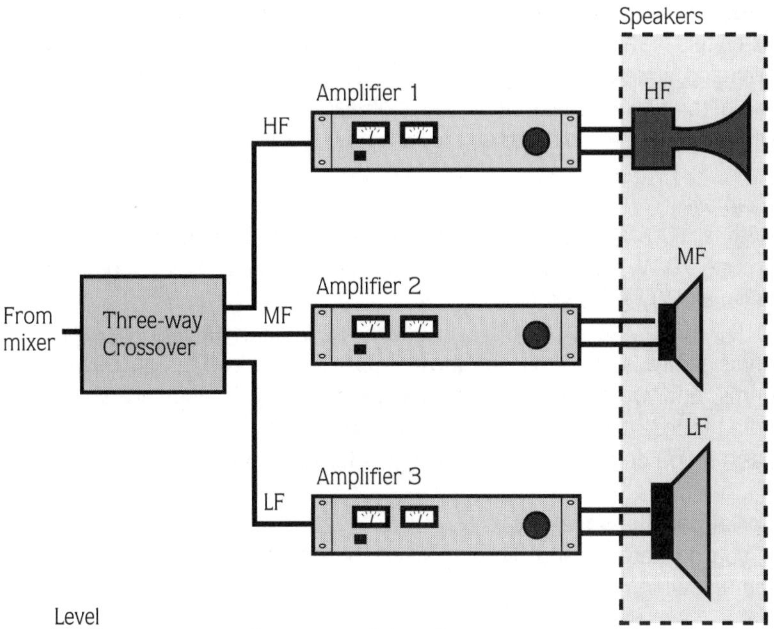

Triamplification splits the amp signal into three bands

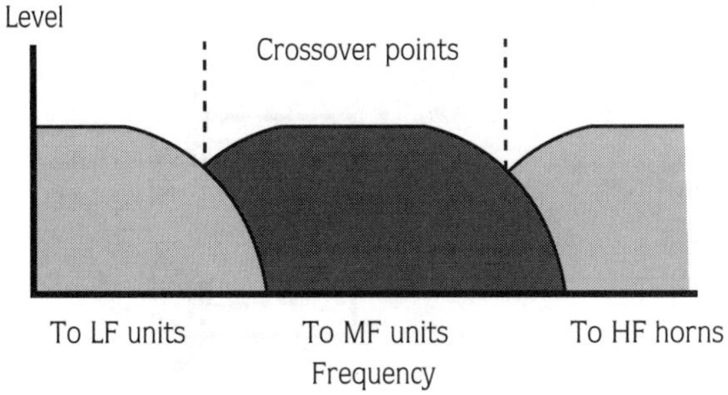

Amplifiers

Each crossover band will need its own amplifier. If we're running stereo these need to be stereo amps, or for mono you can split each stereo amp and use it as two mono ones.

Amplifier/speaker matching

One of the easiest and cheapest ways of getting more power from your system is to actually match the amplifier and speaker impedance properly. Most amplifiers are rated for 4 ohm loads, whereas most speaker units are of 8 ohms impedance. The solution is to run two 8 ohm speakers in parallel off each side of the amp. As well as delivering and transferring maximum power from the amplifier, the combination of speakers working together will increase the SPL by around 30%. The amp will run hotter but it should cope. This is 42% more power without having to buy a bigger amplifier. Wow ! Actually it equates to a 3 dB increase which doesn't sound quite so dynamic, but it is worth having.

It is important to make sure that both the speakers have the same power handling and frequency coverage as they will be fed with identical amounts of power and one may blow otherwise. For a stereo amp, this means we are running four speakers – two on each side. The other benefit of this is that we have extra speakers which we can direct to a specific area and increase coverage as well as just volume. It also means that we are getting slightly less power through each speaker, so that their power rating can be reconsidered as they are now sharing the load.

Example

With a 100 W into 4 ohm amp but using an 8 ohm speaker we get around 70 W. With the 100 W driving two 8 ohm speakers in parallel, we get the 100 W, 50 W in each.

Do not try to extend this technique by adding yet more speakers; 4 ohms really is the limit for most amps. Some claim to go down to 2 ohms, but you need to look at the spec closely to make sure you're not going to lose quality (i.e. lower the damping factor giving a less controlled bass) if you do so. Other amps may just fry driving 2 ohms.

Serial/parallel speakers

If you do need to add more speakers but maintain the impedance, then the following technique is available.

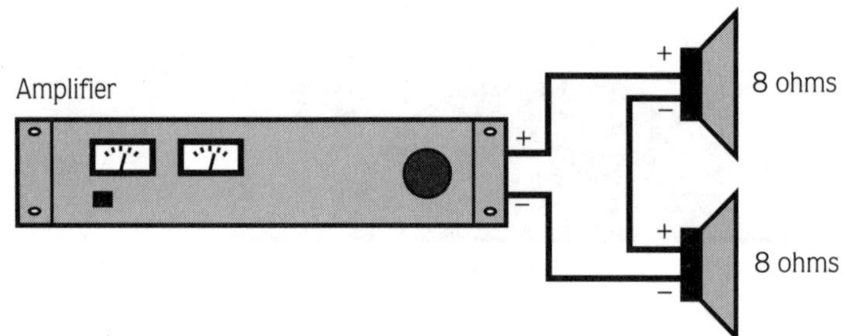

Two 8 ohm loudspeakers connected in series. Connecting in series increases the resistance. The same power is generated in each speaker. The dispersion is likely to be greater as there are more speakers to focus

Basic electronics tells us that if we put two resistances (i.e. speakers) in series with each other, so that the output of one feeds in to the input of the other, and our circuit is connected to the extreme ends of both, then the resistance will rise. It rises according to the formula:

$$\text{series resistance} = R1 + R2 + R3$$

and so on for how ever many resistances. In this case the total resistance will determine the flow of current from the source. If the resistance increases, it will oppose the flow of current more and can be determined from Ohm's law

$$V = I \times R$$

where V is the voltage in volts, I is the current in amps and R the resistance in ohms. If all the resistances are the same, then the power developed across each will be the same. The power developed is provided by the formula;

$$W = I^2 R$$

where W is power in watts, I is current in amps and R is the resistance.

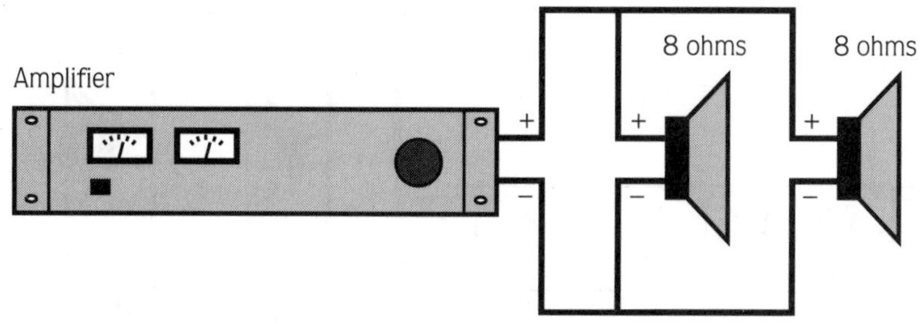

Two 8 ohm loudspeakers connected in parallel. Parallel connection decreases the resistance and can demand more power from the amplifier. The power is shared equally between each similar resistance. Most amplifiers are rated in to a 4 ohm load while speakers themselves are usually 8 ohms – hence two speakers in parallel produce maximum power transfer efficiency.

With a parallel circuit the source feeds all the resistances simultaneously, and the outputs of each resistance are connected together at the other end. In this case the formula for resistance is:

parallel resistance = (R1 x R2 x R3 ...) / (R1 + R2 + R3...)

In practical terms we can see that, if we are dealing with the same resistances (i.e. identical resistance speakers), then the following happens:

If we put two speakers in series the resistance doubles (i.e. 8 + 8 =16). Similarly if we put four speakers in series the resistance will quadruple (i.e. 8 + 8 + 8 + 8 = 32). If we put two speakers in parallel the resistance halves (i.e. 8 x 8 / 8 + 8 = 4). Similarly if we put four speakers in parallel the resistance quarters.

Connecting speakers in series increases the resistance. The same power is generated in each speaker. The dispersion is likely to be greater as there are more speakers to focus. Connecting speakers in parallel decreases the resistance and can demand more power from the amplifier. The power is shared equally between each similar resistance. Most amplifiers are rated in to 4 ohms while speakers are usually 8 ohms – hence two speakers in parallel produce maximum power transfer efficiency.

A common mixed configuration for multiple speaker setups uses two pairs of speakers in series parallel. Note that four *pairs* (eight speakers) of 8 ohms in this configuration would produce 4 ohms to the amplifier.

So if we want to use four 8 ohm speakers, but maintain an 8 ohm impedance, then we can put each pair in series (giving 16 ohms) and then place these pairs in parallel (giving 8 ohms again). In this case the impedance to the amplifier stays the same so it delivers the same power. This power is shared between the speakers and because of this they should have similar power ratings. Although there is no increase in power or SPL, it is a way of using lower power and cheaper drivers and does provide greater directivity and coverage.

TIP

A quick rough guide to calculating parallel resistance is to take the lowest value resistance and to realise that the combined impedance will be less than this. For similar resistances the figure halves for each pair.

Impedance/resistance

The difference between resistance and impedance is that resistance is a measurement of opposition against a DC current such as from a battery, whereas impedance is the measurement of opposition to an AC (alternating) current, such as an audio signal. Impedance measurements hence should be stated with the reference frequency used. In most devices the impedance will be far from a constant with frequency.

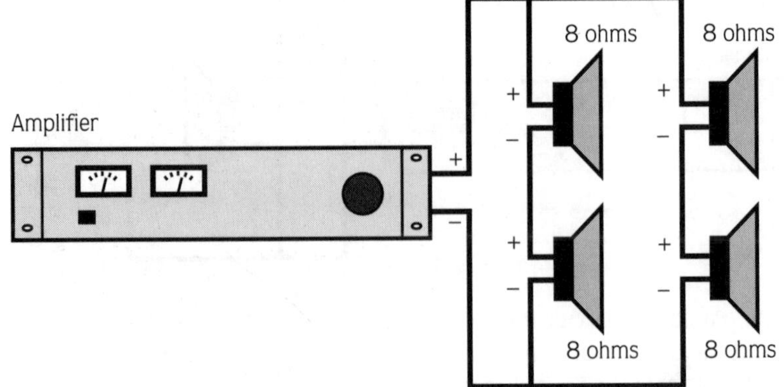

Speakers connected in series/parallel. This is a common mixed configuration for multiple speaker setups. Note that four *pairs* (eight speakers) of 8 ohms in this configuration would produce 4 ohms to the amplifier.

Parallel/serial

If we have two devices we can connect them in two ways – serial or parallel. A series connection means that one feeds in to the other and the same signal flows through both. The power dissipated in each device is related to its resistance.

With a parallel connection, the signal splits in two directions and each path shares the load according to its resistance. The power developed across each load is related to its resistance.

Crossover care !

You must exercise caution when connecting an active crossover system.

1 An active crossover *must* be placed between the mixer and amplifier. It must not go anywhere near the speakers or amplifier outputs as you may well blow it up otherwise. Hopefully the connectors on the system will prevent you from doing this by accident, but exercise care anyway.
2 You must be careful that each output of the active crossover is indeed feeding the correct set of amplifiers and speakers. Feeding bass to an HF driver will most likely blow it up, and feeding HF in to a bass driver will just be very quiet.

Combination mixer amplifiers

The attraction of a combination mixer amplifier is ease of interconnection and setup, as there are less components and connections to deal with. Generally the amplifiers in these are quite small (around 400 W) and the internal effects are often poor. Some of the better units also feature a foldback amplifier showing a much better understanding of what is needed.

The main problem with these units is that of servicing and expandability. If one side of the amp goes down then it is much harder to remedy this quickly. Also when it goes for servicing you lose your whole PA and not just a component of it which you could borrow and still be familiar with.

Expandability is usually a major problem, often allowing no easy way to

Spirit's Powerstation mixer has the added benefit of an integral amplifier and a separate Lexicon effects section offering a range of delay and reverb types

gain more mixer input channels, use better external effects, patch in external effects like compressors, or to up-rate the power amplifier or use a bi-amplification set-up. That is without selling the thing and starting again. This is usually just because of the absence of enough sockets to allow the patching of such devices.

For instance if the internal amp is patchable it could be used for foldback or for one of the ranges when expansion occurs. Some clever combo units allow the user to start his rig using the built in amplifier and then to re-patch it as the system expands to be used as a monitor amplifier. Other systems also have a foldback amplifier in as well, and may allow similar expandability, providing two amplifiers for foldback when expansion occurs.

So it may be easy for now, but when you expand you will probably have to sell it. As you'll lose at least a third of your money when you do so, you need to consider the financial implications of such a system.

System perspective

When talking about a system it is almost impossible not to bring in figures, so let's just put some practical notions to some figures. A dB is just a ratio, usually to a known reference.

In terms of SPL, although 1 dB is in theory the smallest change noticeable, 3 dB is a more practical change. To get a 3 dB increase in SPL we need to double the power of the amplifier (i.e. 100 to 200 W, or 500 to 1000 W). A 10 dB increase in SPL is perceived as being twice as loud. This is 10 times the amplifier power. 0 dB SPL is the threshold of hearing – 0.0002 dyne/cm^2.

System examples

PA system 1 – cabaret

Let's consider some examples. This would be for an average level requirement of around 96 dB which is more cabaret than rock concert. To put it in to perspective, remember that speech is around 70 dB and the threshold of pain is around 120 dB.

room size (in metres) – 10 x 10 x 3 = 300 m^3

Let's say the speakers give 96 dB 1 W !M. We need to cover 10 m , so

20 log (10/1) =20 dB more than 96 = 116 dB
20 dB more than 1 W can be found from W = antilog (dB/10) x Wr
So antilog (20/10) x 1 = 100 W

So what does this tell us ? This shows that a 100 W amplifier coupled with some 96 dB SPL efficient full range speakers, would provide the kind of level we need.

What it doesn't tell us is that we still have the problem that it will be louder for the people nearer the front – the bulldozer effect. Also in such a small venue, it could be very hard to control reflections from surfaces at the rear, so our sound system may be competing very hard with the reverberant field.

In this case the choice of speakers is key. If we use a traditional stack then we need to aim the speakers away from the walls and try to aim the treble units to the rear of the venue, to get some coverage. The speakers also need to be far enough away and aimed to prevent interference between themselves.

The other thing that isn't apparent is that although our average requirement is 96 dB, we really need about 10 dB of headroom, so that any peaks and transients above this will remain clean and undistorted. If we add this to our calculations we get a different result:

Was 116 dB, now add 10 dB headroom = 126 dB which is 30 dB more than 96 dB.
30 dB more than 1 W can be found from W = antilog (dB/10) x Wr
So antilog (30/10) x 1 = 1000 W

So although we won't need a 1000 W of power continuously, we need it to handle the peaks cleanly. This will make a very distinctive improvement in the quality of the system. Remember that the ear detects loudness in terms of contrast, so if the brief peaks are clean, then they form an important part of the equation.

Note this system is aimed at clean sound quality with headroom – do *not* thrash the amp.

The speakers will normally be able to handle brief peaks in the order of five times providing they are clean and of short duration. So the amplifier not distorting is essential. However there is still the risk that the speakers would blow if the amplifier were run at a higher level consistently. So for

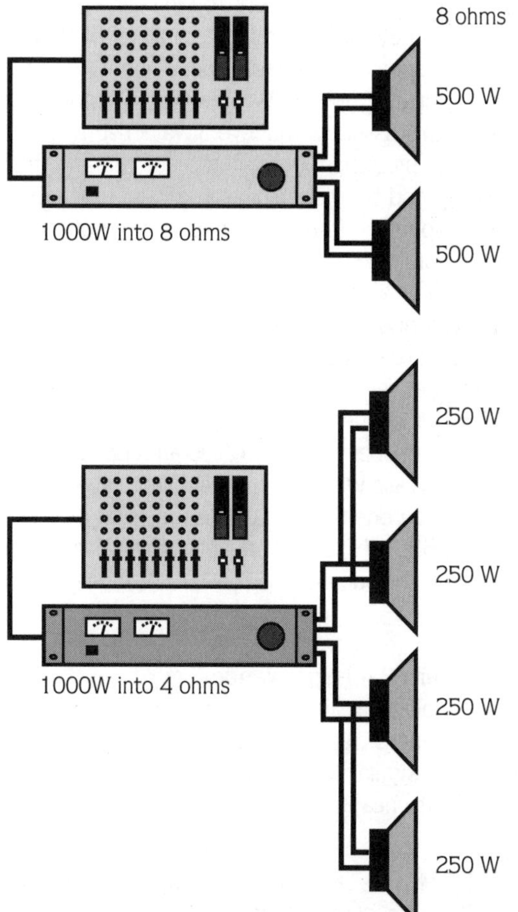

A typical PA system for cabaret use (96 dB average). Note this system is aimed at clean sound quality with headroom – no need to thrash the amp.

optimum safety we should also use larger speakers. However for economy and if common sense is used, then we should be able to get away with speakers in the region of half the power amplifier value, i.e. 500 W speaker for a 1000 W mono amplifier system.

Whether the system is mono or stereo and how we connect the speakers plays a major part in the system. As we are unlikely to be talking mono here, let's apply a twin channel philosophy.

We've said we need a 1000 W. A 1000 W stereo amplifier usually means 500 W per side into 4 ohms. If we use one 8 ohm speaker per side then we run at around 70% efficiency (x 0.7) and only get 350 W per side. We have two options here, we can either use another set of speakers of a similar rating in parallel, which would give us more coverage, or a higher powered amplifier.

If the extra set of speakers we have are not that similar, then we can forego the dubious (discussed later) stereo concept and run each set of speakers off one half of the amp, which would give us a matched set of pairs, where the power of each can be controlled from each half of the amp.

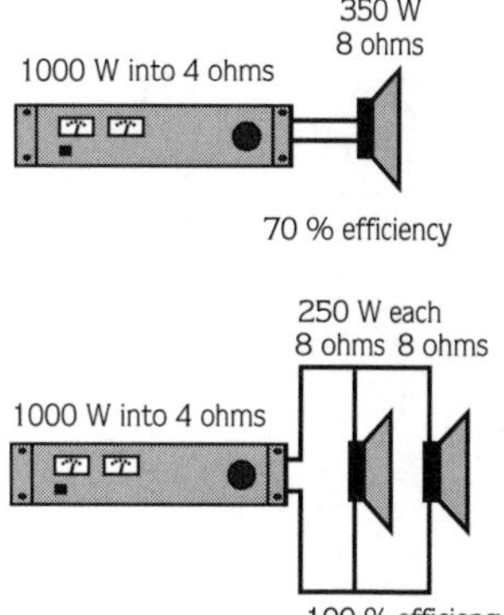

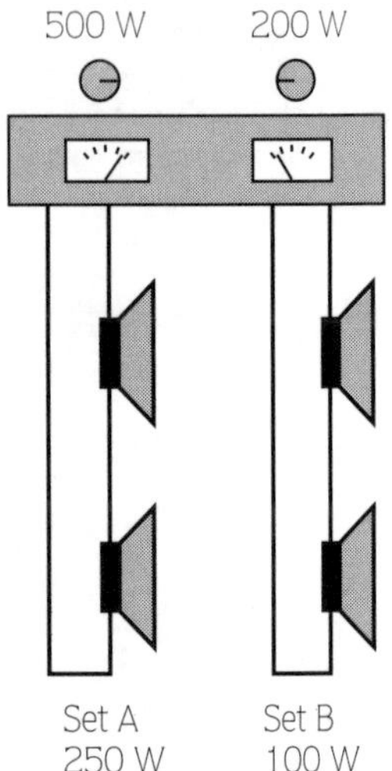

Using a stereo amp in mono to drive different speaker rating types

In this case, as speakers (of the same impedance) in parallel will share the load, they can be reduced to half of 500 W per side, i.e. 250 W each.

If the speakers do have a peak handling of five times, then these could be substituted for 50 W versions, or more practically 100 W ones. Of course this assumes that we always use them correctly – that is remembering to plug them both in and not running the amplifier hot, or the amp may clip and blow the speakers. Remember that the idea is to get a clean quality controlled lower volume sound, not to thrash the amp for raw distorted volume.

There is also the possibility of damage from accidents, like dropping the microphone, which will send any speaker on an excursion of a lifetime, before it potentially dies. We will consider speaker protection later.

Alternatively we can get a higher specification amplifier to get our 500 W back when using an 8 ohm load. This would be (500 / 0.7) around 700 W per side, probably called an XYZ 1400.

In this case we need speakers that can handle 500 W per side, as there will only be one 8 ohm speaker per side of the amplifier. This may ideally mean a 500 W speaker, or again if we use it cautiously, the speakers may be able to be substituted for 100 W types (assuming five times peak handling).

PA system 2 – larger venue

If we look at a slightly larger venue, say 17 x 13 x 5 m, then let's see what happens. Again we want our cabaret type 96 dB and we use speakers of a similar efficiency. Another system aimed at controlled quality sound with headroom – do *not* thrash the amp.

- Note how each amplifier half is used to feed LF and HF only so that the balance between them can be adjusted.
- speakers in parallel for 4 ohms.
- speaker rating assumes five times peak capability.
- directional horns – aimed at rear audience to help cover the space and reduce reflections.
- For a stereo system use twin amplifiers – one for LF and one for HF.

In this case we have to cover a 17 m depth with speakers around 13 m apart. A slightly tall order but let's see what we can do.

Say the speakers give 96 dB 1 W per M. We need to cover 17 m so

20 log (17/1) = 25 dB more than 96 = 121 dB
25 dB more than 1 W can be found from W = antilog (dB/10) * Wr
So antilog (25/10) x 1 = 316 W amplifier.

However again we really want a 10 dB headroom above this, which gives us 35 dB

35 dB more than 1 W can be found from W = antilog (dB/10) x Wr
So antilog (35/10) x 1 = 3100 W

A 3 kW rig is quite a hefty beast, it's also a lot of power to come just from the front of stage, so we really should start to consider multiple placed speakers, reinforcement speakers or at least more focused components at stage front.

Do remember though that we're talking about a quality system running at an average moderate level, so that the peaks are handled nicely. As well as the contrast we get in perceived volume, a clean sound is less tiring and more intelligible, so will be psychologically better than a continuous loud raw buzz.

With this size of venue it is dubious whether one speaker could cover the whole area effectively and we really need to consider using long throw horns with a narrower dispersion so that we can focus them. This will also reduce the reverberant field from surfaces and reduce the bulldozer effect of volume from stage front only.

Also with this power, we should also consider using bi-amplification to reduce the individual component stress and provide a better sound in itself.

Out of interest, if we used speakers of a 93 dB efficiency (3 dB less than 96 dB) then we would have to double our amplifier figures to get that 3 dB back. That could bring us a colossal 6 kW rig. Do remember of course that speaker efficiency isn't the whole story. Speaker response can peak and dip by 10 dB, and it may just be that the laboratory test used a lucky spot. It doesn't matter how loud a speaker is if it doesn't sound good, it will just be bad sounding loud efficient speaker. Your ear has to be the final judge.

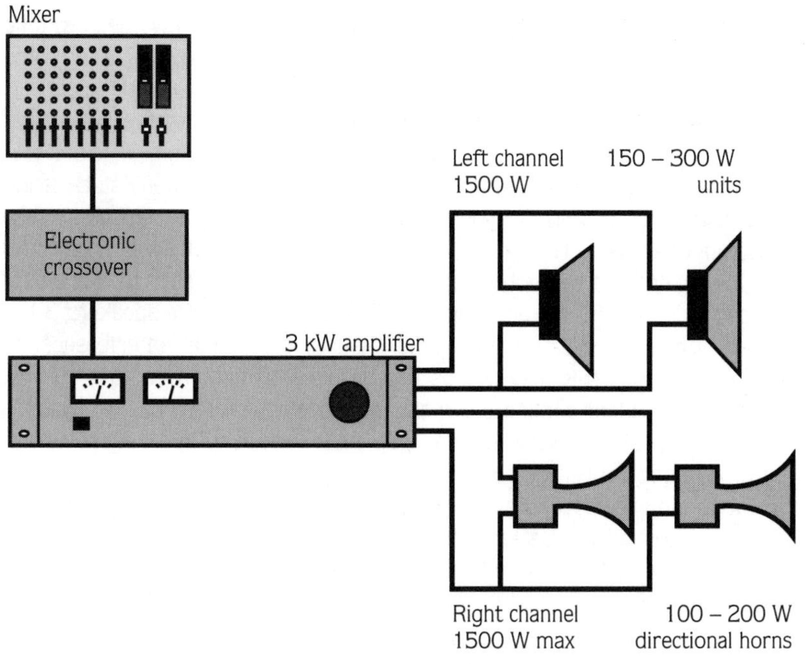

A typical PA system for a larger (17 x 13 x 5 metres) venue. Another system aimed at controlled quality sound with headroom - do not thrash the amp.

Multiple units

If we are going to use multiple speakers, we need to consider a number of factors. Firstly if the speakers are placed coincidentally (in one place) then they tend to work together and the volume of the cabinets add together to effectively become one large cabinet. This is certainly true for the bass anyway. The treble is slightly more complicated. As HF has much shorter wavelengths, the time delays introduced by distance between drivers become important, and it is possible that interference patterns will be introduced as the HF 'comb filters' itself. In such cases we need to take special care about placement (either angling or distance spacing) and choice of component dispersion.

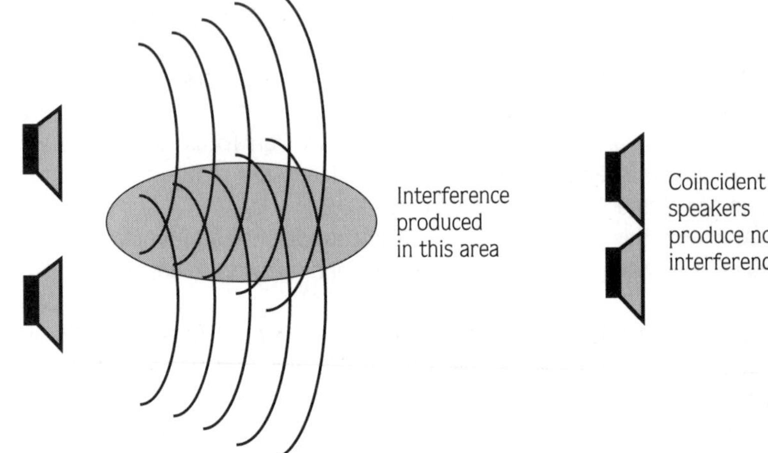

Speaker placement. Assuming the speakers are supposed to deliver identical mono material, coincident speakers behave as one source. With more spacing phase interference may result.

With greater spacing (below) each speaker is fairly independent also reducing the phase interference effects. Speakers should either be coincident or far apart i.e. follow the mic 3:1 rule.

10 m

30 m

If we separate the cabinets by small distances, the interference effects are likely to become worse as well as becoming a problem at lower frequencies. As a starting guide you should try the coincident stack using slight angling to improve your chances, or use a spacing technique (like microphones) of 3:1. That is, the speakers should be at least three times as far away from each other as they are from the audience they are supposed to cover. So for an audience 10 m away, space the speakers 30 m apart. This will greatly reduce the chances of speaker interference. In many situations this may not be practicable, certainly in on-stage terms, and the coincident angling technique will have to be used. But for systems involving speakers spaced over a wide area, such as with ceiling speakers, it gives a good coverage guide.

Assuming the speakers are supposed to deliver identical mono material, with the coincident speakers they behave as one source. With more spacing phase interference may result. With greater spacing each speaker is fairly independent, this also reduces the phase interference effects.

Bi-amplification and tri-amplification

We mentioned earlier that we should consider bi-amplification in our system, so let's have another look.

Bi-amplification means splitting the signal into two frequency ranges (bands) and using equipment to handle each band separately. As well as reducing the stress on each component it provides a cleaner sound as each component can concentrate on a specific job. It requires a separate amplifier for each band.

Bi-amplification systems use an active electronic crossover which is placed between the mixer and the power amplifiers. *It goes nowhere near the speakers.* A typical crossover frequency is 800 Hz, and the following chart shows options for tri-amped systems as well.

Bi-amp and tri-amping crossover frequencies

	Crossover point	
Band	*2 way system (bi-amp)*	*3 way system (tri-amp)*
HF		2.5 kHz – 20 kHz
MF	800 Hz – 20 kHz	800 Hz – 2.5 kHz
LF	40 Hz – 800 Hz	40 Hz – 800 Hz

Bi-amplification will reduce the chances of interference between the frequency ranges and will also allow optimum power to be delivered to each band, reducing any clipping and overdriving effects. If you are using compression on the outputs of the crossover, it will also help in terms of frequency masking any compression effects. Bi-amplification also reduces the risk of blowing speakers as any clipping distortion that occurs will be confined to that band. So a bass clip will not be sent through the tweeters, and conversely the bass driver won't waste its time trying to reproduce any HF. As such it will tend to give a smoother response and sound.

As HF is 3 – 5 times more efficient than LF, we can use amplifiers of the required power range. So with a 1000 W amplifier to handle the bass, we can use a more economical 200 W amplifier for the horns.

Although you might consider such a system as only a 200 W PA, it is really still a 1 kW PA. It is human hearing and the physics involved in moving volumes of air at bass wavelengths that produce these HF LF efficiency discrepancies, and these are just the proportions required.

Sub bass

Passive crossovers may still be used in any sub bass bins which will deal with the 40 Hz and below range. This range tends to be felt more than heard, and can be important in some musical styles. These sub bass frequencies should be separated from normal bass speakers if at all possible, as the amplifier and bass speaker will waste valuable power in trying (and failing) to recreate these very low frequencies. The result will be less defined regular bass response. If sub bass bins are not available, you might do well to filter out these low frequencies using the mixer's high pass filters. Sub bass may also be the culprit for feedback which builds up due to room excitation and reinforcement of standing waves. Again filtering will help to avoid such problems.

Critical distance

So now we have a feel for what is happening at the speaker front, we need to look closer at the rear of the audience. Not so much as in assessing the backs of their heads or their haircut, but in terms of what they are hearing in the cheap seats.

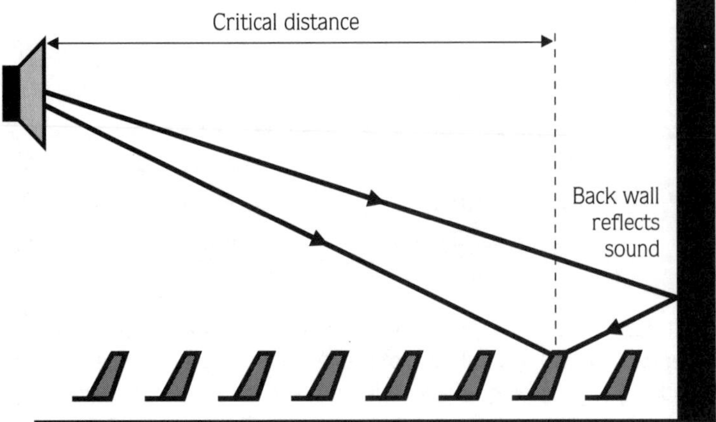

Critical distance in speakers. At the critical distance the reflected sound will compete with the direct sound and produce poor intelligibility and mush. This is often around the 7 m distance. Reflected sound is loud enough to obscure the direct sound – poor intelligibility results

In a field (or an anechoic chamber – a padded acoustic cell) the sound you hear is direct. There is nothing for the sound to hit and reflect back to you. That is why for outside venues we need more speakers and more power, but have less quality problems (see later).

In contrast a venue has many reflective surfaces, including floor, ceiling and walls, which can really mess up our sound. Because the reflections are time delayed and the frequency content of them is complex (different surfaces absorb and reflect different frequencies) causing coloration, the resultant combination of direct and reflected sound is far from perfect.

At more than 7 m (21 feet) from the direct source (the speaker) the reflections become a large factor. This is called the critical distance and is the point where the direct and reflected sounds become equal in volume.

There are only two ways to avoid this problem. Firstly to change the acoustics of the venue and reduce the reflections. But this would need to be done at all frequencies, not just treble, so curtains won't help here. Or if we could make the reflected sound cleaner (much like making the pick-up spill from a microphone pleasant) and re-use it in a useful way then that would also be a solution. But this would bring us in the realms of acoustic design which is beyond the scope of this book.

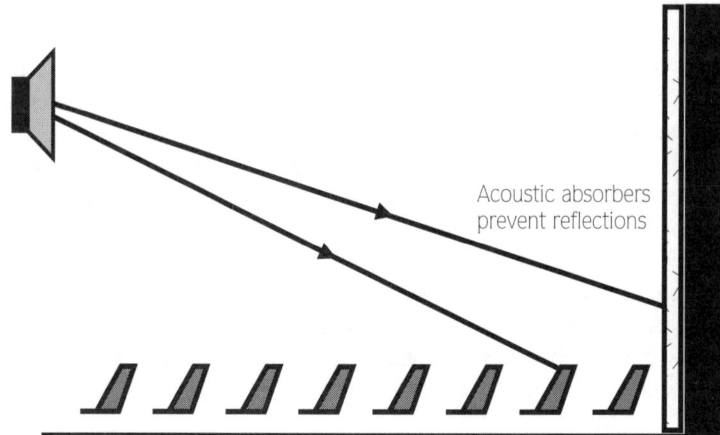

How to increase the critical distance:
(a) Reduce reflections through use of acoustics and screen absorbers.
(b) Angle the speakers at the audience and avoid surface reflections.

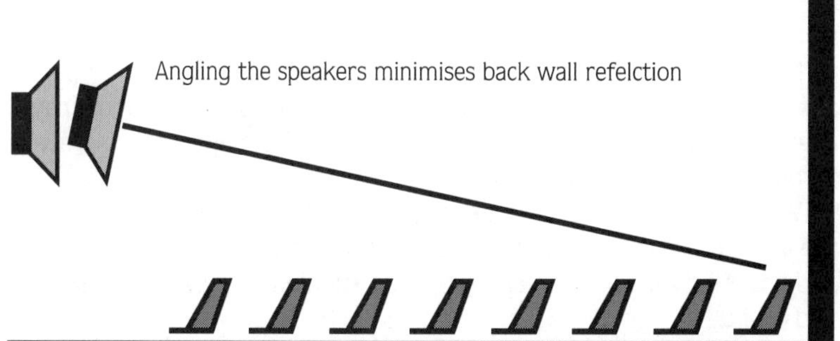

The second solution is to use speakers with a narrower dispersion and more carefully aimed at the audience rather than the surfaces. This is where the concept of compression drivers comes in.

Compression drivers

The compression driver basically uses a horn arrangement to focus the energy from the speaker into a narrow dispersion but long throw device. This has been compared to the garden hose. With a wide nozzle it covers a large area with water and the range is limited. With the nozzle tight, the coverage area is much smaller, but the pressure has increased and the distance possible is much greater with a larger volume of water reaching a smaller area – i.e. more wetness per square inch.

The compression horn does the same thing acoustically for sound. The horn starts from a small throat which extends via a flared channel. The sides of the horn focus the sound and basically determine the dispersion area. The amount of compression depends on the size of the throat and the length and opening of the horn. Because of this, compression drivers tend to be quite a bit larger than a regular horn – up to 18 inch flare for an HF unit.

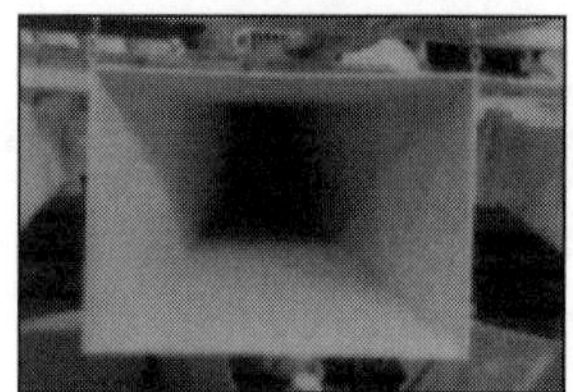

The horn starts from a small throat which extends via a flared channel. The sides of the horn focus the sound and basically determine the dispersion area (Pic courtesy Shuttlesound)

There are a number of choices in dispersion with the compression driver to allow the ideal focusing regardless of the distance. For the nearer field, a wider dispersion compression driver is used, and as we go further to the rear, narrower dispersion drivers are used. As always, some care has to be exercised to avoid interference between the horns' projection patterns, but luckily this can be done visually by following the eyeline of the straight part of the horn, and verified by soloing each horn and using ear or testing equipment measurements to confirm the response.

When setting up the system it should be remembered that the acoustics of the venue will not stay constant. For example a venue has a lot of HF absorbing material (called the audience) which appears after the sound check. This needs to be compensated for on the fly as the audience don't really like listening to pink noise, test tones or being moved out of the way for a sound level meter.

Outdoor venues

Outdoor venues have a mind of their own. Because of the lack of surface reflections and sound containment, we should consider using 2 W per person (in the area to be covered) as our guide. Also remember that they are likely to be spread over a much wider area and may even be travelling through your sound stage at whim.

Wind and rain are other random factors which will affect your system, the latter one being a problem of protecting the gear from water damage, and the first one being responsible for running off with your sound in mid-flight.

Multiple lower level speakers are advised here and fortunately time delays are not as much of a problem in the open (it's direct rather than reflected sound). If time delay compensation is required, sound travels at around 0.3 m/ms. As we noted earlier, use a slightly longer delay than that required to maintain the distance perspective.

Obviously great care needs to be taken with outdoor events in terms of cable laying, equipment harnessing and security. With the audience moving so freely it can be a major problem.

Mono vs stereo

The concept of stereo for a live venue is a highly debatable one. Although the sense of 3D perspective is desirable for any sound system, the realities of live sound are that most of the audience will not be in any sort of position to benefit from the effect, especially those at the fringes and those who are further way from the speakers than the distance between the speakers.

Also avoiding larger than life stereo images is usually accepted, with the concept of a 7 m wide drum kit being an example. So extreme static panning is probably to be avoided. What does work well however is occasional dramatic panning, such as of sound effects which occur occasionally, and the stereo sense of travel would be a useful addition to the sonic effect. However any mainstay instruments will not benefit from such an effect. Although the keyboard may be on the left, or the string sound desired from the left, the fact is that for the audience on the other side of the stage, it will just be constantly quieter. At the extreme edges it may in fact be too quiet. So if any panning is to be used, it should be subtle, unless it is transitory in nature anyway.

Stereo reverb and delays can be used to quite good effect though and can be panned widely. This is because they don't carry the main parts of the performance so no important information will be lost. However the effect will be a useful indication of the spread of the sound, and if not heard partially by an element of the audience, will not be a major loss.

The use of stereo equipment should not be precluded. In fact it is hard to buy mono equipment these days, even single speakers can be hard to purchase. Using stereo equipment as twin channel devices is the option.

By splitting a stereo amp, one side for LF and the other for HF, gives us independent control of each band and gives us the start of a bi-amplified system. Or one side for FOH and one for monitors. Similarly any true stereo effects units can be used as two independent mono devices, by ensuring that any link or stereo mode switches are bypassed.

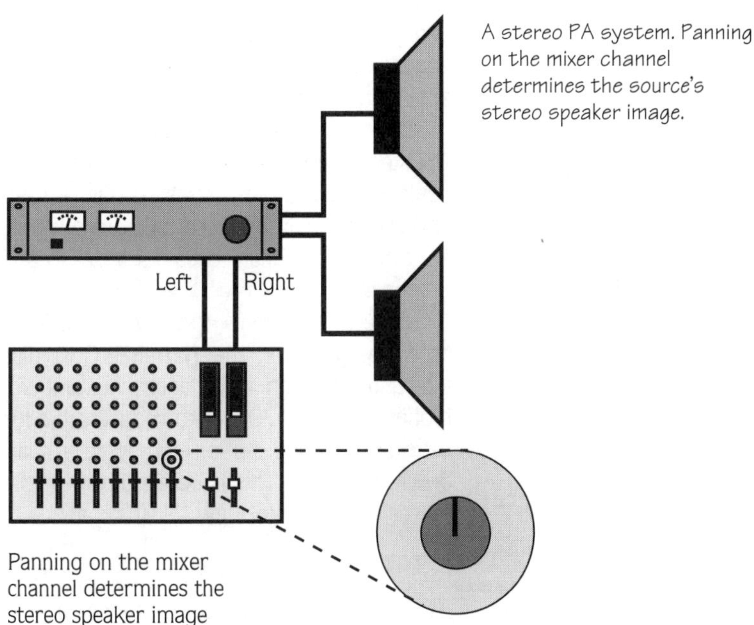

A stereo PA system. Panning on the mixer channel determines the source's stereo speaker image.

Bridged mode

Some amplifiers feature a bridged mode, where special connections can be used to provide an amplifier of double the power of its stereo mode. This is a useful option.

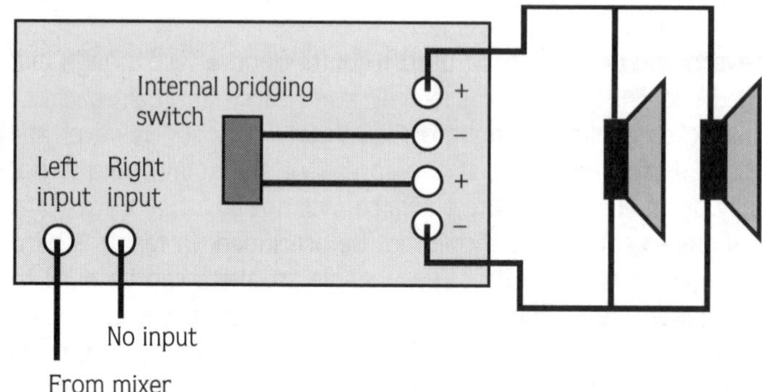

Bridged mode amplification. Bridged mode changes a stereo amplifier into a mono amplifier of twice the power. Take special care to follow the manufacturer's connecting instructions! Some amps may sum both inputs, others may use just the left channel input.

100 volt line

For applications where the need is for a modest spread of sound, multiple small speakers of low power can be placed regularly throughout the venue. This system is ideal for conference rooms and pubs or strange shaped venues (such as L shaped rooms). The speakers need to be spaced around at least 3:1 (speaker:audience) apart to stop interference.

The other factor is that with a system using 50 speakers, we could use 25 stereo amplifiers to drive them. This is obviously impracticable. We could use a complex series parallel arrangement to distribute the power and retain our impedance, but there is still the problem of miles of speaker cable which can cause severe loss of power and high frequency response.

Modern amplifiers (post valve) use low impedance constant current type designs. For the National Grid distribution system the electricity board use a high impedance constant voltage type system, using much higher voltages (20 kV) to reduce cable losses, which is why you probably have an LEB sub station somewhere near you to bring the voltage back down to 240 V. This is because power loss in cables is due to heat caused by resistance. Ohm's law tells us that

power = current squared x resistance

As we can't reduce the resistance of the cable much, by increasing the voltage and reducing the current (but maintaining the delivered power which also equals current multiplied by resistance) heat loss through the cable is reduced.

The system for speaker installations to do this is called 100 V line. The term stems from valve amplifier days, which were naturally high voltage high resistance designs. These days we can still use transformers on regular transistor/mosfet amplifiers to create a 100 V line system.

The 100 V system relies on the fact that the amplifier is transformed into a high impedance constant voltage type amplifier using high voltage.

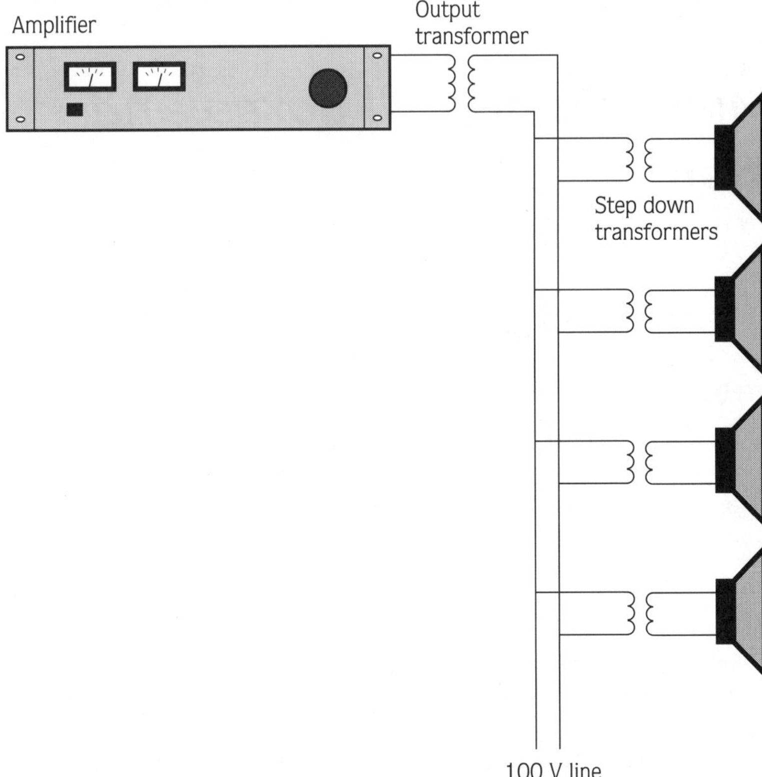

A 100V line system. The 100V line output transformer, which must be very high quality, is usually built in to the special 100V line version of the amplifer – but it can be added. It converts the low impedance of the amplifier into a high impedance constant voltage source so that the impedance of the multiple speakers becomes less prevalent. The 100 V line system is intended for multiple *low* power speaker configurations (i.e. under 30 W each) over a large area – using a high impedance transmission line system. Each speaker is fed from a volume tapped transformer to match the impedance of the speaker back.

This causes less heat loss in the cable. Then on each speaker, a 'step down' transformer is used to return the signal back to the low impedance current based level required by the speaker. As the transformers for the speakers are quite low powered, the system isn't as expensive or bulky as you might imagine.

This system is ideal for venues and conference areas where multiple speakers are required to reinforce the sound, rather than provide it. You've most probably experienced it at train stations, pubs and restaurants. It is also a useful system for exhibition and museums where a local quality audio feed is needed rather than the bulldozer blasting of the rock stack.

4

Microphones, DI boxes and more

Intro

No PA system would be complete without a means of capturing the sound. Here we look at microphones and live sound microphone techniques, wireless systems, DI boxes, splitters and sub mixers.

Microphones

Talking in analogy mode for a moment, some people judge a person's wealth by the quality of his shoes. Apparently this is the thing that most people forget to upgrade until they have lots of money. In PA terms I suppose this would equate to microphones, because people seem happy to spend thousands on their amps and speakers and then begrudge paying more than forty quid for a microphone.

A microphone's job is to turn acoustic energy (vibration of air) in to an analogous electrical signal which the mixer can understand. One can draw a lot more parallels with the camera rather than with the ear.

Microphone types

It is customary to look at microphone construction when discussing microphone applications and we'll make no exception here. In practical terms it is only helpful to know about construction in terms of original choice and to understand some of the physical problems in using each type. So here we go.

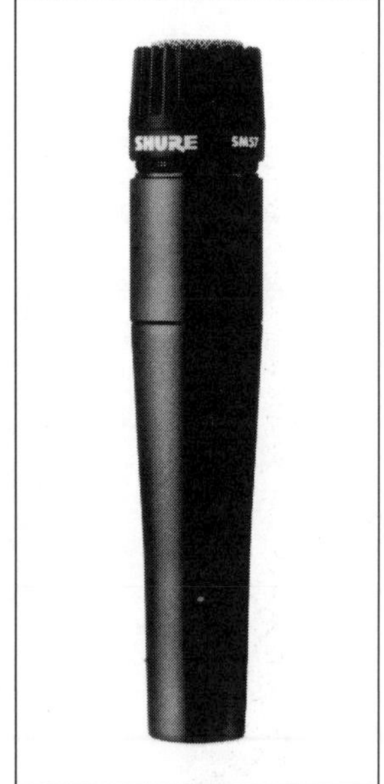

The Shure SM57 – a well regarded live instrument mic

Moving coil mics

The most common microphone is the so called moving coil type. It is very much the reverse of a loudspeaker, in that a coil of wire, suspended in a magnetic field, is connected to a diaphragm cone. When the air vibrates the diaphragm and attached coil, a current is induced in the coil which is analogous to the sound.

Moving coil microphones are good because they are general purpose and can handle high levels; they are cheap, robust and moderately good sounding. The most expensive one is probably the Electrovoice RE20 at around £535. The cost effective point for a moving coil is around £120. Their quality depends upon diaphragm weight, rigidity and suspension. In

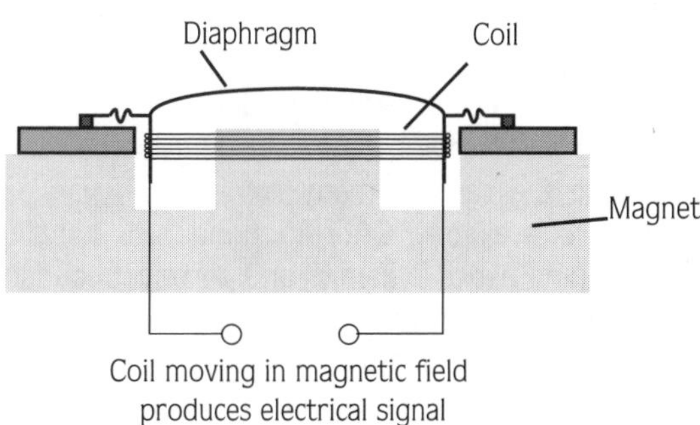

Moving coil mics rely on sound waves moving a diaphragm which has a coil attached. The movement produces an electrical current in the coil because it is suspended in a magnetic field.

practical terms they are good sounding general purpose microphones which can take quite a lot of abuse.

Capacitor mics

The capacitor microphone comes under a number of different names – capacitor, condenser and electret. These names stem from their original design and powering system, but my preferred term for all of them is capacitor mics, as that is the technology they use.

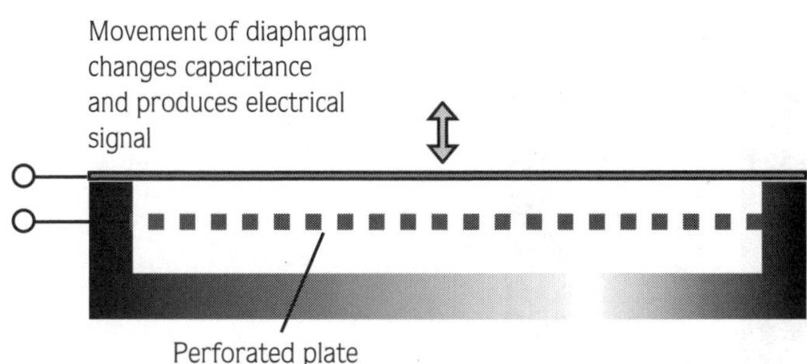

In a capacitor mic movement of the diaphragm produces a change in the capacitance between two plates. To be effective the plates need to be charged

In essence the capacitor microphone consists of two plates – one is fixed and charged and the other forms the diaphragm. When the air moves the diaphragm plate, it causes a change in an electronic characteristic called (not surprisingly) the capacitance.

These microphones tend to do exceptionally well at reproducing bright transient sounds like percussion. They are also excellent on snares, as overheads and on acoustic guitars. They are not generally ideal for very high SPLs such as kick drums. They are more fragile than moving coil types and are prone to problems with wind and moisture.

Using a capacitor microphone without a windshield (one may be built in) will collect moisture from the mouth which will make it produce a frying crackling noise. This is easily remedied by placing the microphone under a lamp or spotlight for a while, but that's not much use to you in the middle of a gig!

The AKG C1000 is an example of a budget capacitor mic intended for vocals (built in windshield) at around £230. The cost effective point is around £250, and the most expensive range from the famous Neuman

The AKG C1000 – an affordable capacitor mic

U87 (not really a live microphone though) at around £1200 to the Calrec Soundfield at £2800. This uses capacitor capsules and sophisticated electronics for post event variable positioning.

Moving coil mics are the most popular type for live sound use, but this needn't be the case as many robust road versions of capacitor microphones are available. Offering good SPL handling, excellent transient response (and hence dynamics and perceived loudness), extended frequency response for high frequency clarity and high output level, they can be ideally suited to the live sound situation. Phantom power is available on a wide range of mixers or can be supplied by separate free standing units. With normal installations and cables, there is no need for phantom power to cause concern to a PA system.

Ribbon mics

The ribbon microphone was so named because it consists of a very thin aluminium ribbon, thinner than bacofoil (chicken oven wrapping) suspended in a magnetic field. Because the ribbon is so thin and quite large, it is excellent at capturing the nuance of a sound and is bright without being clinical, as some capacitor microphones can be.

Ribbon mics have a very fine ribbon which moves under the influence of sound waves.

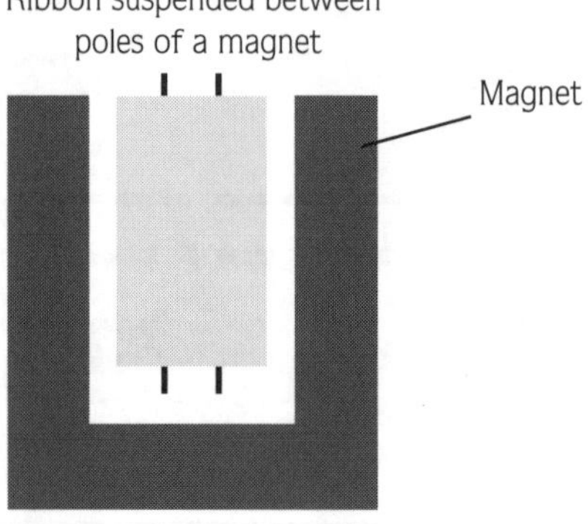

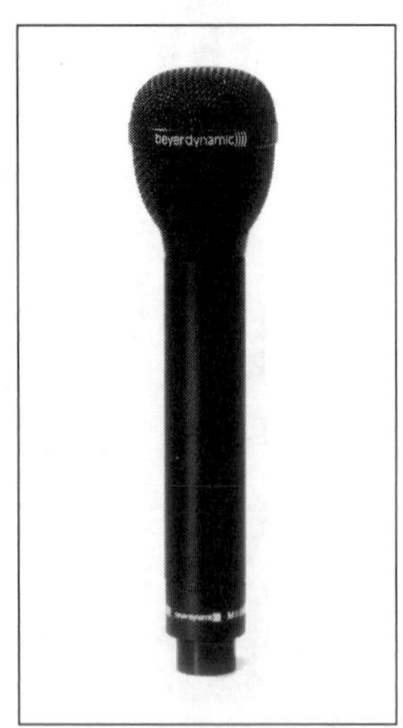
Beyer's M88 is a rare example of a robust ribbon mic

They are excellent for strings, emotional vocals and as overheads. They tend to suffer from handling noise and aren't ideal for mobile boom arm use, as in TV, or for outdoor work, where wind noise and rumble may be prevalent.

For live sound, specifically designed examples include the Beyer M88 and M260. The cost effective point is around £200.

Polar patterns and microphone housings

The housing of the microphone capsule plays a major part in its polar characteristic (response to frequency with direction) and frequency response. Following the laws of physics, microphones with larger capsules tend to have a naturally low frequency response (hence the AKG D12 and D112 and Electrovoice RE20 for kick drums), although many microphones have been tailored to produce similar results despite their size. How well they manage to do this without colouring the sound or transient response is for personal debate.

The main thing about microphone housings, is that there are two types – pressure operated and pressure gradient. These are very similar concepts to the speaker enclosure sealed and vented designs.

Pressure operated microphones

Pressure operated microphones consist of a capsule which is open to air pressure in only one direction (the diaphragm). This gives it a natural omnidirectional response, making it equally receptive to sound from all around. This design offers the advantages of design simplicity and cost, lower handling and pop noise, and no proximity effect or coloured response affected by distance.

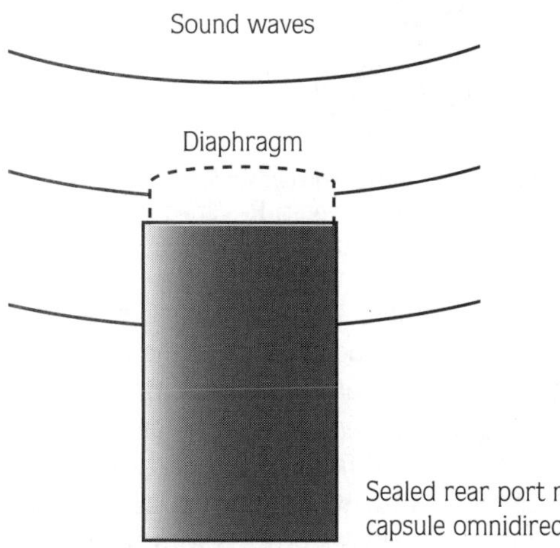

In pressure operated mics a sealed rear makes the capsule omnidirectional

Because of these facts, they make ideal microphones but are not really suited for the live sound environment, when feedback and spill rejection are normally paramount. However, if you can afford to use an omnidirectional mic as an overhead in terms of gain before feedback, it will produce very good results.

Pressure gradient operated microphones

Pressure gradient operated microphones have both sides of the diaphragm exposed to air. This gives them a naturally figure of eight response, as is common with ribbon microphones. They reject sound coming from the sides and are useful for dual pick up, such as with mid conga placement, toms to toms, and toms to cymbals. They are also quite popular for radio play and interview work.

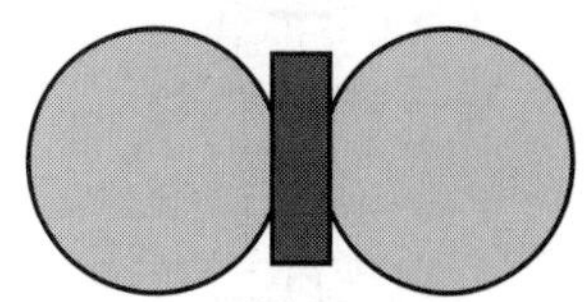

Figure of eight response

Pressure combinations

The microphones we see today tend to be combinations of these two techniques, offering a rear path which is restricted by a designed and controlled set of paths. This provides us with our classic cardioid (heart shaped) response picking up sound mainly from the front, a little from the sides and very little from the rear.

Pressure gradient mics have an acoustically designed rear port path which allows front and rear phase cancellation producing a directional response.

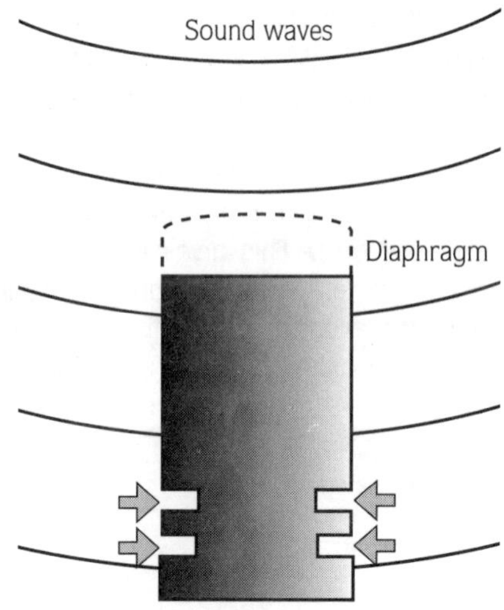

Acoustically designed rear port allows sound waves to enter. Resulting phase cancellation causing directivity

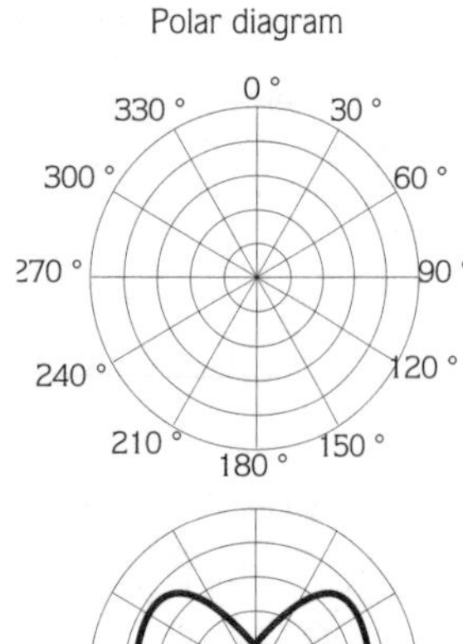

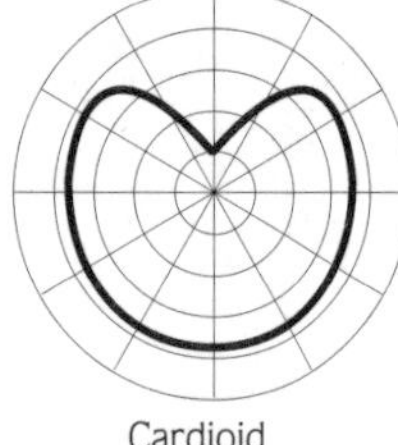
Cardioid

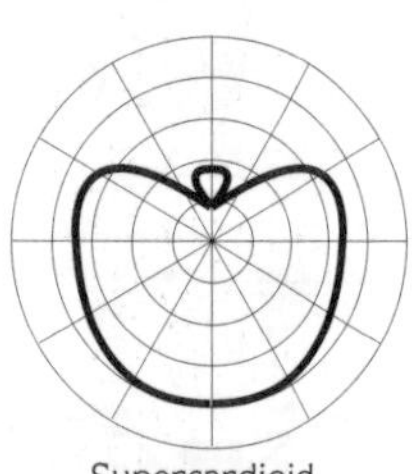
Supercardioid

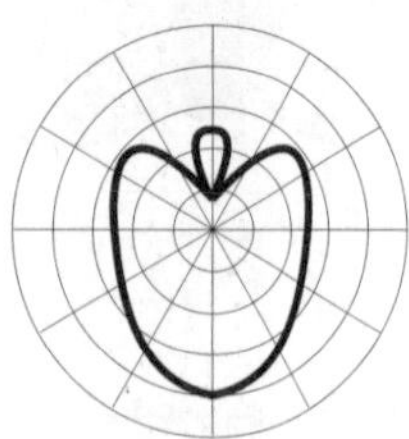
Hypercardioid

Microphone polar pickup patterns

For sound coming from the front, the diaphragm can move back, meanwhile sound refracting around the rear and entering the rear paths, arrives out of phase and hence strengthen the flow of the diaphragm. Conversely sound arriving from the rear is designed to arrive out of phase between the back and front of the diaphragm and hence cancel.

The situation with sound arriving at the sides of the cardioid microphone works on a similar process with some cancellation and reduction in volume occurring.

As with most things audio, the polar response is very dependent on frequency: bass tends to be omnidirectional while high frequencies are more directional. This is due to the effective size of the capsule and paths compared to the wavelengths of the frequencies involved.

Microphone polar pickup patterns

The major categories are as follows:

- Cardioid – heart shaped pickup from front and sides, and rejects from rear
- Supercardioid – tighter sides heart shaped pickup, but with a rear pickup lobe
- Hypercardioid – even tighter heart shaped pick up but with a pronounced rear pickup lobe.
- Figure of 8 – front and rear pickup with rejection of sides. Useful in live work for toms.
- Omnidirectional – even pickup from all directions (excluding mic body masking)
- Hemispherical (half globe) – accepts all front boundary sound, but rejects rear of boundary sources.

Proximity effect

In fact this gives rise to a phenomenon called the proximity effect. This is an increase in low frequency response when used very close to the source. This is again due to the fact that the path to the rear of the microphone is very short and insignificant to bass frequencies and the phase difference at low frequencies is negligible, causing a boosting to occur at these low frequencies.

This can be a useful tool if a bass heavy warm sound is required and is a useful chance for gain before feedback at these frequencies. However the effect is very pronounced with distance, and a movement of less than a couple of centimetres can cause a drastic change in response. For this reason these microphones need to be maintained at a consistent spacing to the source.

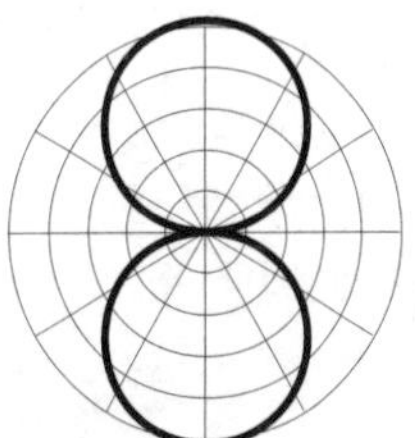

Figure of eight

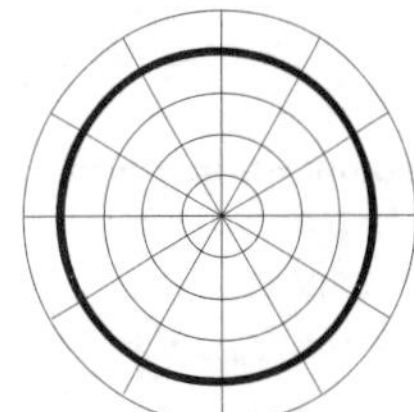

Omnidirectional

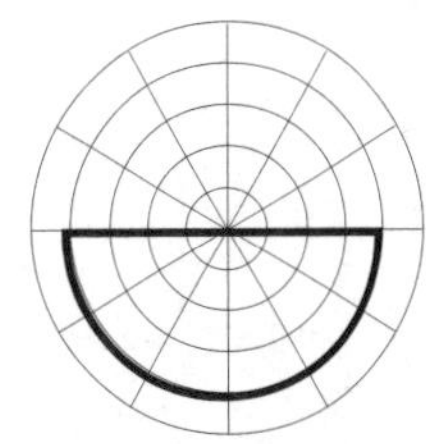

Hemispherical

Hand cupping

You may have noticed the effect of feedback howl if hands are cupped over a microphone. This occurs for two reasons: first, by blocking the rear vents, the microphone can become omnidirectional and pick up more sound, and secondly because the hands act like a horn and focus the sound energy on part of the microphone, causing an increase in gain. The practical point of this is to urge people not to cup their hands over the microphone as all you can do in such a case is to turn it down, or kill them and hope they let go before *rigor mortis* sets in.

Polar response vs coloration

In a studio situation, spill is not so much of an issue as having usable spill which doesn't spoil the sound. Although rejection and separation are desired, a good sound is preferred. Of course in a studio you can do retakes and use screens. For live sound there are no such luxuries!

It is a variable situation as to whether general enhancement of the stage sound is sufficient, or whether engineer control is necessary to give a good sound. This is mainly dependent on how well balanced the sound is naturally, or whether egos are a performance factor ('I'm louder than you are' spirals).

In PA terms we tend to stick to cardioid microphones. Supercardioid and hypercardioid microphones are also available. These offer even less side pickup and a narrower front pickup. However the penalty for this is an increase in pickup at the rear, in the shape of a lobe. With very tight microphone spacing requirements, it might be better to use straight cardioids as the lobe may provide even less desirable coloration than the spill. Otherwise they are a good choice.

Some capacitor microphones have pickup patterns that are selectable by using a clever system of two capsules and phase reversal techniques. These however are not normally found in live use.

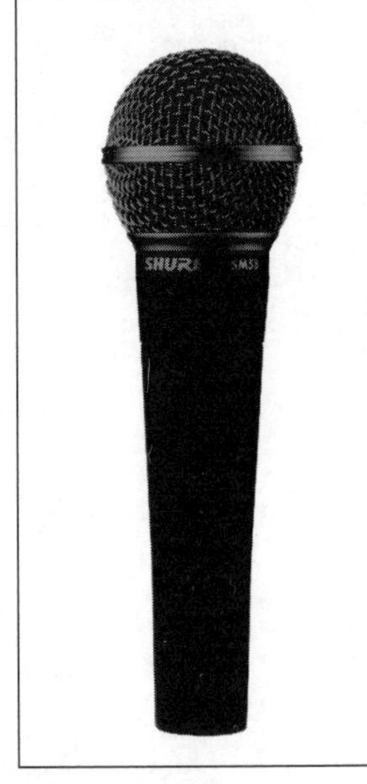

The Shure SM58 – an industry workhorse

Microphone applications

Construction	*Cost*	*Robustness*	*Quality*	*O/P*	*Applications*	*Comments*
Moving coil	£120	excellent	average	average	general, overheads/ambient	proximity effect/ pops
Capacitor	£200	fair	excellent	high	percussion, vocals (use windshield)	suffers from moisture
Ribbon	£250	delicate*	excellent	low	acoustics, strings	prone to high SPL, shock and wind

Typical microphone usage

Instrument	Models	Type
Vocals, brass and combo amplifiers	Shure SM58	m
	Shure Beta 58	m
	Shure SM87	c
	Electrovoice N/D 457	m
	Electrovoice N/D 857	m
	AKG C1000	c
	AKG D3700	m
	Beyer M88	r
	Beyer M300	m
	Sennheiser 427	m
	Sennheiser 421	m
Acoustic guitar	Shure SM57	m
	Shure SM81	c
	Beyer M201	m
	AKG C451	c
Piano	Shure SM57	m
	Shure Beta57	m
	Shure SM81	c
	PZM	c
Reed	Beyer M201	m
	Shure SM81	c
Kick drums	AKG D12	m
	AKG D112	m
	Electrovoice RE20	m
	Beyer M88	r
	Sennheiser MD421	m

Typical microphone usage (cont)

Instrument	Models	Type
Snare	Shure SM57	m
	Beyer M201	m
	AKG C451	c
Hats	AKG C451	c
Toms	Shure SM57	m
Overhead	AKG C451	c
	AKG C414	c

* Some manufacturers do design ribbon microphones which are robust. For instance the Beyer M88 is a ribbon microphone which is recommended for use with bass drum as well as vocals.

Microphone frequency response and coloration

Choosing a microphone on specification alone is almost pointless. These days, nearly all microphones have a wide frequency response and can handle the levels involved. All the ones we'd consider for live use will be some variation of cardioid polar response anyway. Four things dictate the choice of a microphone:

1. the sound – judged by ear
2. the useable level or sensitivity of the microphone (meaning savings in mixer gain)
3. the flatness of the microphone frequency response or any desirable peaks or dips and
4. the rejection of feedback and low handling noise

These factors can really only be judged empirically – by trying them in anger. We can of course shortlist them by some criteria, but application also depends on the type of music you handle, the artists you work with, your PA system and the venues you play. These are things the microphone manufacturers can't know.

As a very personal guide, here is a view of the character of the different brands so that you know which ones to definitely check out. I stress that this is a very personal view of the characteristic sound of these devices and some models may contradict this fingerprint. However, armed with this brief you should find it easier to choose for yourself.

Mic characteristics

	Warmth	*Power*	*Sharp*	*Thin*
AKG	***	**	**	***
Beyer	*****	***	*	*
Electrovoice	****	****	**	*
Sennheiser	*	*	***	****
Shure	**	**	**	*

The logic behind microphone choice is that it should contrast with the source. In the case of vocalists, if the singer has a thin voice, a warmer mic would be more likely to complement it. Conversely a vocalist with a warm voice would probably benefit from a thinner crisper sounding microphone. The same logic can be applied to the PA system itself once you've decided what its bias is.

Personally I find it is better to go for different brands in different areas – so for instance if I was using Roland keyboards, I would use a Yamaha reverb. The same seems to apply to microphones and speakers. So I would tend to choose alternate brands of microphones to my speakers. Speakers also have their own characteristic sound and again contrast is advised to provide an overall balance.

It will be appreciated that a choice of microphones to cope with any situation would be a benefit, but a complete alternate set isn't always financially possible, so you need to choose equipment that matches your own sonic fingerprint.

Gain before feedback

We've spoken a number of times about gain before feedback. This concept springs from the fact that the more un-natural gain is introduced in a system, the more likely it is to feed back. Therefore the higher level the microphone and speakers can produce, the less mixer and amplifier gain is needed, which is good news. Similar principles are true of large equalisation correction and coloration with the system itself, which will increase the chance of feedback at resonant frequencies. We'll discuss feedback in more detail later.

Specialist microphones

For live music work, we have traditionally used the regular microphone and boom stand. In broadcast and conference work other microphone types are already popular.

Lavalier

The lavalier (or lapel mic) is a very small microphone which can be attached to the person or an instrument. Apart from the trailing cable, it makes for a very unobtrusive and perhaps more importantly, mobile microphone system, as when the source moves, the microphone moves with it. This negates any pickup problems or problems of sound with distance, as is the case with microphones exhibiting proximity effect (extreme bass tip up at close distances). By adding a radio system to the lavalier microphone, we really have a mobile system.

Shotgun mics

In some live applications it can be advantageous to use distant miking methods, such as with shotgun (rifle) mics. Like the telescopic lens on a camera, this allows us to zoom in on a source from far away. In situations where the performer insists on being mobile and radio mics are not available, creating a hot area using shotgun mics is certainly an option. In live terms it will tend to offer better results in terms of feedback and sound control, to creating a hot area with overhead or floor mics, or PZM zones, as the hot spot can be made much more selective. The spill from shotgun mics can sometimes be rather strange sounding, so using them as additional fills rather than the main focus can be advisable.

PZM (pressure zone microphone)

The PZM mic (sometimes called a boundary mic) isn't so much a specialist microphone, as a conventional microphone in a special housing. The principle is that by mounting the capsule very close to a boundary, the sound arriving at the capsule from directed and reflected sources will be almost identical, and all sound forms part of the pressure zone which the microphone picks up. This means that the boundary becomes part of the microphone. These mics usually come on small 150 mm (6 inch) plates, but the size of the plate can easily be expanded by mounting it to a surface such as a floor, wall or screen. These can even be clear perspex (or similar) for less visual clutter.

The pickup pattern of a PZM mic is hemispherical – it rejects sounds from behind the boundary and accepts sounds from 180 degrees in front of the boundary with almost equal response. This makes it more difficult to use in a live environment but not impossible. We can control the pickup by using V shaped boundaries and even mounting a PZM on each side of a panel and using it similar to a figure of eight response.

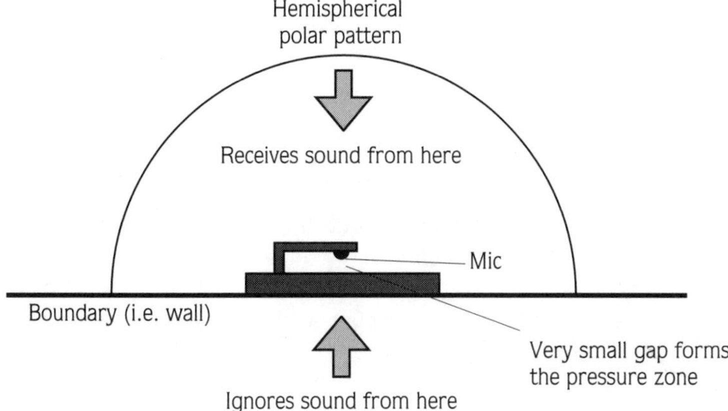

Pressure zone mics turn the mounting surface into part of the mic

Mounting methods for PZMs (boundary affects bass response)

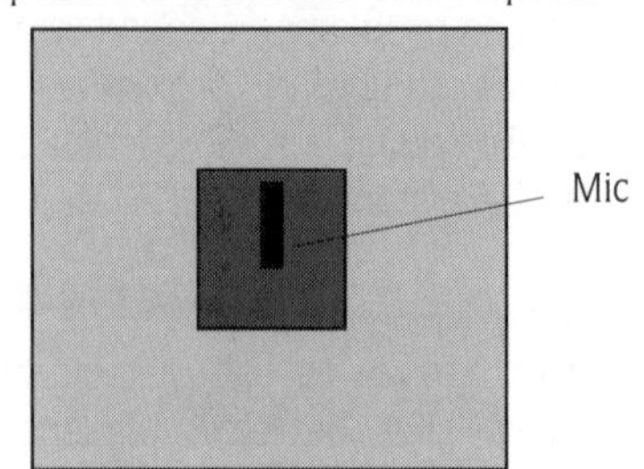

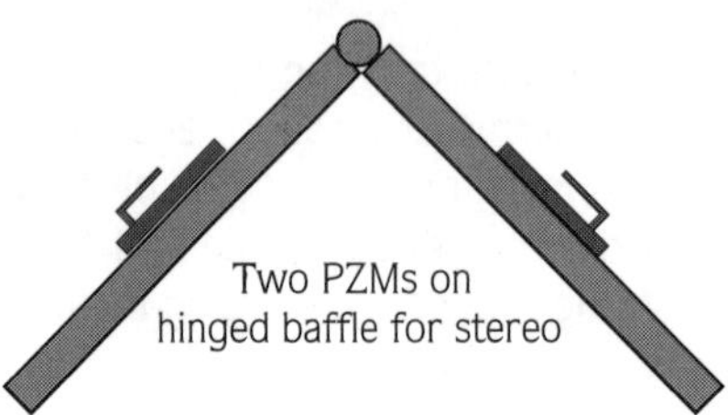

> **TIP**
>
> *The bass response of the PZM mic will be determined by the size of the boundary, so at least 1.2 m square is advised for a healthy bass response. The boundary size needs to be around a quarter wavelength (C=WF, 100 Hz has a 3 m wavelength so 750 mm would suffice for this, or 1.5 m for 50Hz).*

PZMs excel at picking up the overall ambient sound. This is why they are ideal for conference work, where the entire lectern becomes the mic without introducing the psychological intimidation of phallic microphones. It also gives the speaker more opportunity to move about without affecting the sound so drastically. In live work they can still be very useful if feedback can be controlled, and they will reduce the chances of interference and balancing problems that occur with multiple mic techniques.

It is common for PZMs to use miniature capacitor capsules (requiring phantom power), although this is by no means a necessity. Some PZM microphones are slightly noisy and many of the cheaper ones offer a very variable quality between units and batches. So watch out for these factors before you buy one. Some companies (like Electrovoice) do a housing to convert any mic into a PZM type if you want an alternative.

Choosing a microphone

Firstly there are *no* rules about microphone applications. What works for one person one day may be totally inappropriate on another. You can only judge the success of a microphone by ear.

The only guidelines are physical ones:

1. Don't use a ribbon microphone for heavy percussion like kick drums, as they may get damaged
2. Microphones with integral windshields (mesh balls) are usually intended for vocal applications
3. Microphones with flat heads/grills (rather than balls) are meant as instrument microphones and may not work well as vocal mics (i.e. pops and moisture frying)

4 Capacitor mics tend to offer a better HF and transient response but may not offer the smoothest sound.
5 Choosing a microphone with the sound you want, rather than equalising it, will usually offer better results.
6 Use similar mics for all the vocals to make equalisation against feedback easier – it can be done globally on the main output graphic rather than corrected on each channel.
7 Moving a microphone a few centimetres can drastically change the sound. This luxury of experimentation isn't very common for live sound, so experiment before the show.

In PA we tend to have to go for the loudest part of the instrument rather than the most characterful, but it is about getting the sound we want as well as just getting a loud sound.

Microphone techniques

Again there are no hard and fast rules here – only microphone common sense. Microphone common sense is really about understanding where on the instrument each part of the sound is coming from.

Yes, the sound of an instrument is comprised of radiation from many parts of its surface. Even the human voice is made of resonances in the chest and nose, not just the throat. By using close miking techniques (usually to avoid feedback and spill) we are making a frequency selective decision about the character and sound of the instrument.

It may sound obvious but you need to listen to the instrument with your ear and even move your ear around to hear how it sounds. In studio situations it is common to send this mic down the foldback so the engineer can hear the results of mic positioning as he does it. This technique could be transferred to the stage if you set up an engineer headphone feed. You could even borrow the drummers click track feed if he has one.

In the studio we have the luxury of being able to use distance to achieve the desired sound. In PA terms 'angle' is one of our few friends. We can angle the microphone without compromising the distance and attempt to get the sound this way. This gives us far more opportunities to get a more balanced sound and to refine it. This is another place where the directional pickup of the microphone can be used to our advantage. In other words we can use angling across the instrument, whilst still retaining rejection against spill and feedback in the opposite direction.

Multiple mics

The alternative to this of course is to use extra microphones to capture the whole instrument. In PA terms this is more difficult than with recording, because of the closeness of the stage and the problems with spill, feedback and microphone interference.

When placing multiple mics, go either for coincident techniques, where the microphones are very close together but angled away from each other, or use the 3:1 spacing rule. The 3:1 spacing rule means the microphones should be three times apart from each other as they are from the source. This, like the coincident technique, reduces the chance of interference problems between the microphones.

Microphone separation should be at least three times the distance from the user (or they should be be coincident) to avoid interference

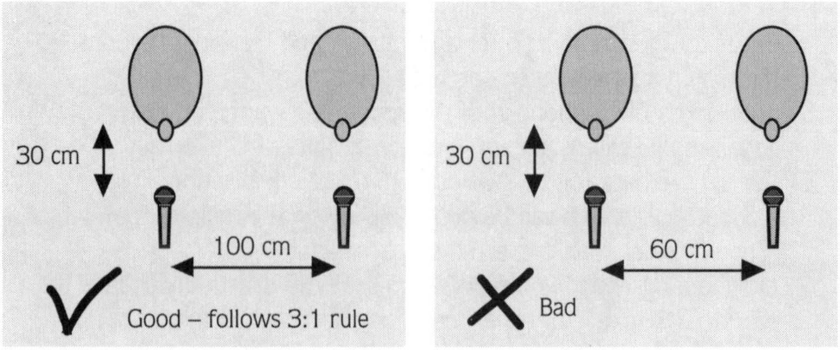

30 cm
100 cm
Good – follows 3:1 rule

30 cm
60 cm
Bad

By angling the cardiod mics we can reduce the 3:1 rule

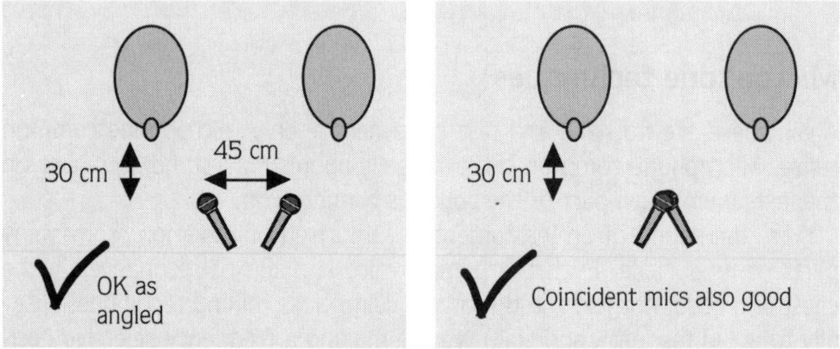

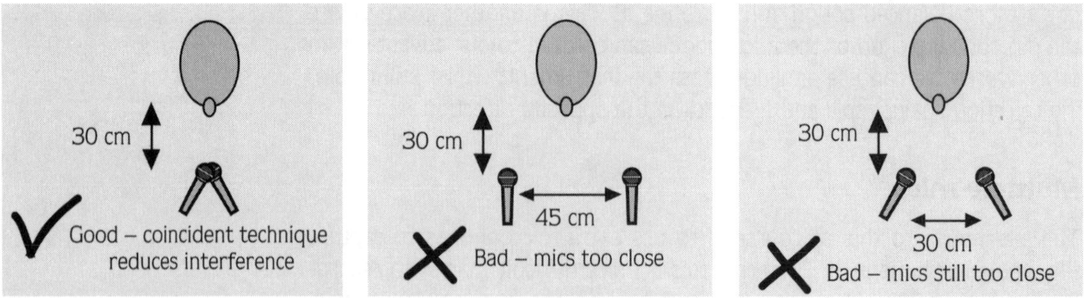

30 cm
45 cm
OK as angled

30 cm
Coincident mics also good

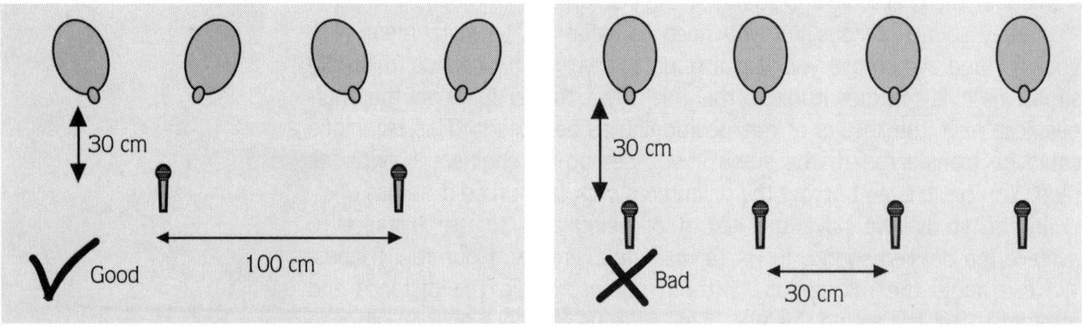

30 cm
100 cm
Good

30 cm
30 cm
Bad

30 cm
Good – coincident technique reduces interference

30 cm
45 cm
Bad – mics too close

30 cm
30 cm
Bad – mics still too close

Double miking of single sources is normally done only for safety (news conferences etc.). But if both signals are combined then follow the 3:1 rule

When miking sources from opposite sides, such as a top snare mic and underneath (snare side) mic, remember that the two microphones will naturally be out of phase, as one skin is being pushed away with the stick, whilst the other skin is being moved towards the underneath mic. In such cases it is important to reverse the phase of the underneath snare, or a lack of bass will result when the two signals are added together.

Stereo microphone techniques

Where massive volume isn't such an issue a number of stereo microphone techniques can be tried. As these tend to be ambient techniques, they are not suitable for high volume work owing to the risk of feedback. For filling or ambient level increase they can be ideal.

They are particularly useful in the reverse role for capturing the all important audience ambience needed for live recording, and where the signal isn't fed to the speakers on the night (i.e. just to the recording).

Coincident XY

This is the technique where two directional (i.e. cardioid) microphones are placed close together, pointing at around 90 degrees to each other. As they are close together, phasing problems do not occur and the arrangement offers good mono compatibility. This system tends to offer a good stereo image but with a poor centre image.

Spaced XY

By separating the microphones by less than 30 cm, we get a better stereo image but mono compatibility isn't so good due to phasing effects. But used in conjunction with close mic techniques, can add depth.

Spaced omni

By placing two omnidirectional mics, several feet apart, we get a good stereo image and mono compatibility, but at the expense of a hole in the middle effect. This can be filled with close or spot microphones.

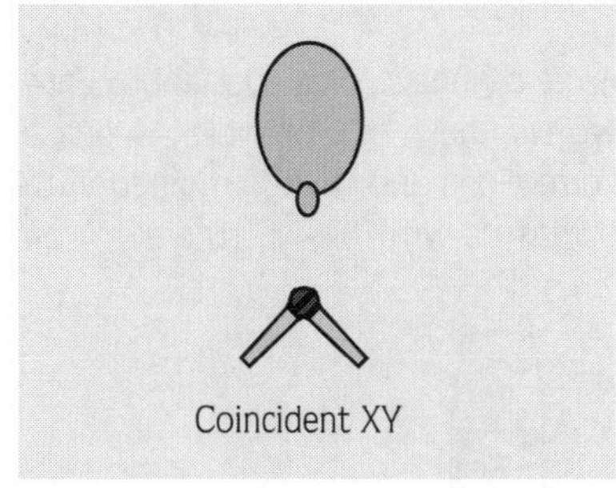

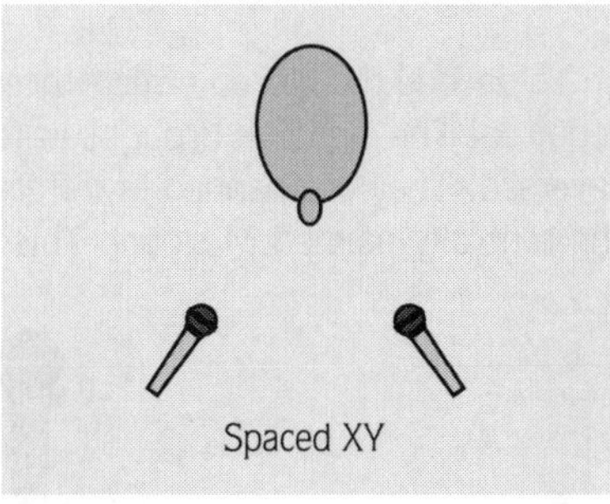

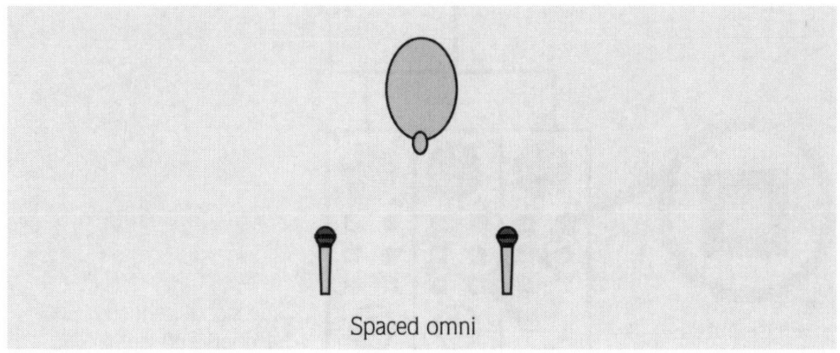

Stereo microphone placing
(a) Coincident XY. Acceptance angle is 90 – 180 degrees. Provides good phase response with well defined stereo edges but less defined centre image
(b) Spaced XY gives a better stereo image but mono compatibility isn't so good due to phasing effects
(c) Spaced omni gives good stereo image and mono compatibility, but at the expense of a hole in the middle effect

MS – Middle/side (sum and difference)

MS is an interesting technique which allows some flexibility in imaging. It revolves around using a central microphone of any polar response to pick up the middle (M) signal. A figure of eight microphone is then used at a coincident point to deal with the side (S) signals. Mono compatibility is excellent as it relies on the one central microphone.

MS stereo – ideal mono compatibility as mono signal is present. Also allows decoding flexibility (i.e. width). Provides a good centre image but with less defined stereo edges

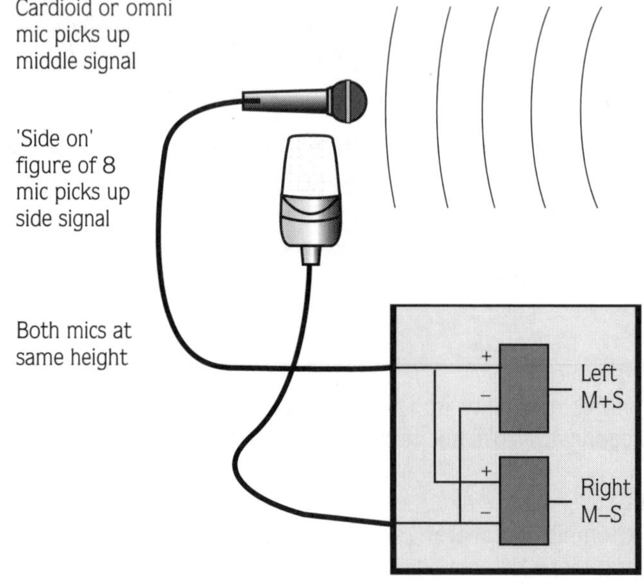

Sum and difference circuitry

To decode the stereo effect, three mixer channels (or a special box) are required. The signal is brought in to two channels, one of which is phase reversed. They are panned in different directions and then combined with the centrally panned M signal. This provides us with M + S (the sum) on

MS decoding

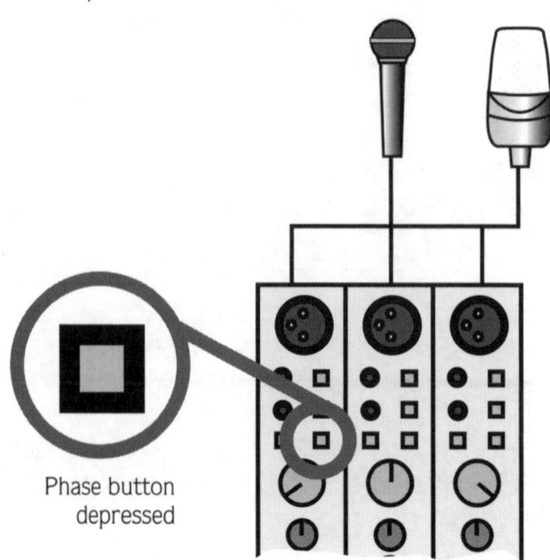

one side (left) and M – S (the difference due to the phase reversal) on the other side (right). By varying the levels of these signals, we can control the width of the image.

For recording, this decoding technique could be left till the mix if the two signals were left separate.

Identification

One of the biggest problems in live sound, is knowing which microphone is being used. Performers have a tendency to use the wrong mic and swap positions which can lead to nightmares trying to locate the correct channel. The microphones should be easily identifiable from a distance with the use of coloured windshields, coloured tape, or coloured cables to distinguish them.

Taping the cables down with gaffa tape can stop them being swapped. Then, if you have followed some kind of logical plan, like running the mixer channels in the same direction and order as the microphones stage left to right, you have some chance of fading the correct mic up at the required time.

Radio mics

Cables are the bain of everyone's life. They get tangled, they are never long enough and are a mess. The radio mic is the potential angel to rectify all this, allowing performers to roam the stage without the microphone leash. In practice radio mics offer almost as many problems, if not more, and these are invisible radio wave wires we're dealing with now !

Types of radio mic

Radio mics come in three main packages; *integrated with a mic* (bulky hand held but no need for a belt pack), *component microphone* (allowing any microphone to be used but being a belt pack with cable to the microphone) and *component instrument* intended for guitarists and the like.

Using a radio mic

With any radio system the biggest bugbears are the batteries needed to power the remote bit. These are traditionally bulky and the transmitter pack often clips to a belt while a small cable runs to the traditional mic. On these systems you can use any dynamic mic you like, which is excellent for flexibility. Capacitor mics needing phantom power can not be used. Some systems have a 5 – 9 V supply available from their own battery to power electrets but that's about all.

The receiver is a mains powered unit with an aerial and an output jack on it. Traditionally these go near the mixer, so that the engineer can check the receive strength meter and adjust the level and squelch controls. It also means the output from the receiver which is at unbalanced line level, can go straight in to a line input on the desk, rather than through a DI box and down yards of multicore cable.

The Trantec radio mic system consists of mic unit, belt pack and twin aerial diversity receiver

However in venues with problem reception it can be an excellent idea to place the receiver near the stage box (on stage) so that there is less distance involved. Wireless mics use FM wavelengths which are essentially line of sight, so anything you can do to reduce reflections from pillars and other obstacles will help. It also mean you have less chance of radio break up or breakthrough, as the signal will be stronger. As it's line of sight, you might find raising the height of the receiver improves reception.

Most modern wireless systems use a thing called a diversity system. These means they have two aerials and will automatically switch to whichever is the strongest. This makes quite a difference for stable sound in most venues and is highly recommended. It means that whatever the obstacle, at least one signal will find its way through.

It is essential to check the stage area for any dead spots before the gig. As the artist is mobile there will be more feedback opportunities, so you need to define any hot or dead areas with them before hand.

When choosing a radio mic system, remember that there are a limited number of frequencies available and it isn't uncommon to find other devices using the same frequencies. Since the DTI tightened this up a little while ago this has improved from the days of picking up most cab companies in the area, but security paging systems and the like are still out there waiting.

Testing obviously helps to uncover these problems, but they often arise at odd times of day (or night!). For instance your CB man may only go on after nine when his kids are asleep and you're well in to your gig. The only precautions are to have spare frequencies as wide apart as possible and to have a regular cable mic set up as a backup. Some sound is preferable to a performer with no sound.

Radio mic golden rule: Change the battery for a fresh (alkaline) set before the performance and use the half dead ones for the rehearsals. Battery life is unpredictable at the best of times and it's a silence you can't afford.

Are you legal?

Since the DTI tightened up in 1993, it became illegal to sell or *use* a non approved system, and fines up to £20,000 can be imposed. As the onus is also on the user now, you would be wise to double check your system. Most of the branded systems have now converted, although some were incredibly slow to do so, so it is only the cheaper imported systems that are still circulating. Remember that ignorance is no defence where the law is concerned – and I've told you now anyway. Get legal !

DI boxes

For most PA work, microphone sources are used. However keyboard feeds, radio mics and direct outputs from guitar systems are not. The majority of stage boxes only feature microphone level balanced inputs so a means of connecting these line sources to them is required. Do NOT use a lead converter ! You run a serious risk of blowing up the equipment due to the 50 V phantom power coming down the mic cable. Also the level and impedance will not offer the best results sonically. The humble DI (direct injection) box will convert the sockets and electrical conditions for you and protect your performer's gear from phantom power.

Passive or active?

There are two type of DI box – passive and active. The passive type uses a transformer to do all of the work. These offer good results for the price but are not perfect as they rely on a passive transformer, which is just a load of wire on a coil. Depending on how well they are made they can introduce distortion and loss of high frequency response. But most are very acceptable.

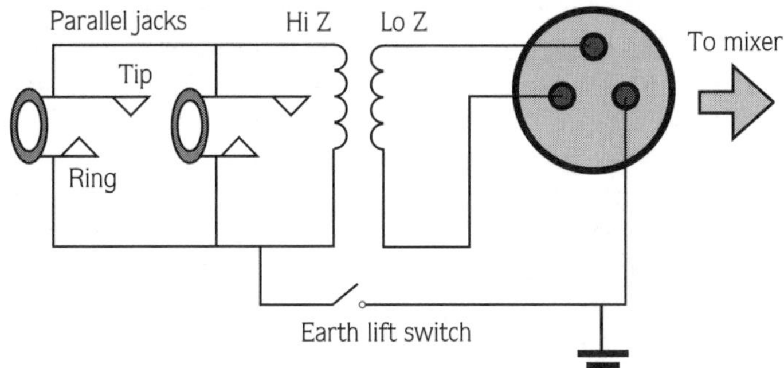

Passive DI box circuit

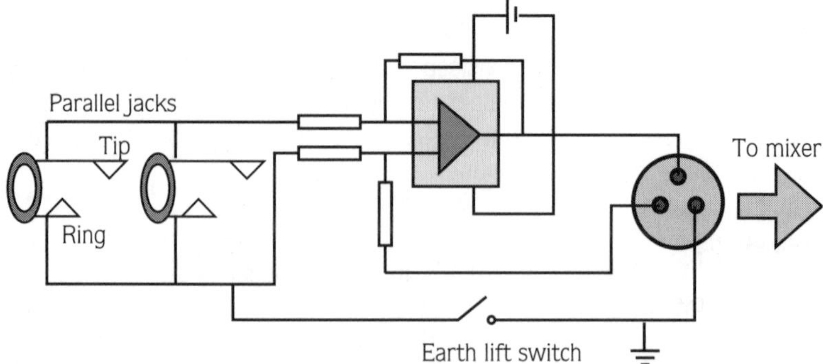

Active DI box offers a powered electronic solution

> **TIP**
>
> A very common reason for no level out of the DI box is that the high level inputs have been used by mistake instead of the instrument input. If it's very quiet, check it out.

The active DI box offers a powered electronic solution, much like the mixer circuitry itself. However a poorly designed electronic one will do much more harm to your signal than a good transformer one, so follow your budget carefully. The better active DI boxes can be battery or phantom powered, the latter saving an awful lot of grief and money. It also means that they have the chance to run off 48 V rather than 9 – 18 V, providing the opportunity for more headroom.

DI boxes usually feature an earth lift switch. This is very necessary to cope with the different grounding conditions at a venue. Sometimes we want the earths extended to the mixer, other times we don't. The test is to flick it and see if it hums any less. If it is quieter then leave it that way.

Some DI boxes also offer extra sockets in addition to the conventional instrument in, parallel link (to continue to the performer's local amp) and microphone output to the stagebox. Some also feature high level inputs such as from extension speaker outputs of amps. This allows you to get the sound of the pre-amp, without having to mic up the speaker. Most amps rely more on the speaker sound than the electronics, but it is a useful compromise. Often you will use both a mic and this amp output. Do not connect high level or speaker outputs in to any socket on the DI box, other than this one.

Splitters

When working with a separate monitor mixer or live recording or broadcast team, it is usual to split the existing mics rather than double rig them. Double rigging offers greater freedom for choice and serves as a backup, but there are enough things to go wrong already, let alone the mess of extra cables.

If you are going to share mic feeds, you need to do this as soon as you can in the signal chain. You don't want another engineer's gain setting to distort your signal do you? To get two outputs from the microphone, use a signal splitter. These are often twin output transformers (or they can be electronic). In essence they provide a buffered feed and compensate for impedance mismatch problems. Unfortunately they soon add up in cost, so just hope the other company has got them (joke!).

Sub mixing

Although not strictly an input device, taking a feed off a sub mixer can be a useful option. For instance, keyboard sub mixers can save lots of extra cables and mixer channels. This works fine if the keyboard player knows what he's doing and can balance himself. At least he knows when his solo is coming up.

There are a number of places to take this feed from, including main outs, tape recording feeds, group outputs, an auxiliary, direct channel outputs, or you can use the insert points (channel or group) if you re-link the ring and tip connection after tapping off. The insert sends are usually pre-fader post EQ so you would be relying on him getting the coarse gain and EQ correct.

A similar idea can be used to feed a separate monitor mixer if microphone splitters are not available.

A typical sub mixing system uses two mixers to create more inputs by feeding the output of one mixer into two spare channels of the other. Two channels are needed for stereo use. To share the foldback and effects busses, use two extra channels as shown – but remember to keep faders down as shown. Use pre-fade auxiliaries for foldback, and post fade auxiliaries for effects sends. On main mixer keep auxiliaries down. Keep faders down. Aux links are used to share main effects and foldback system

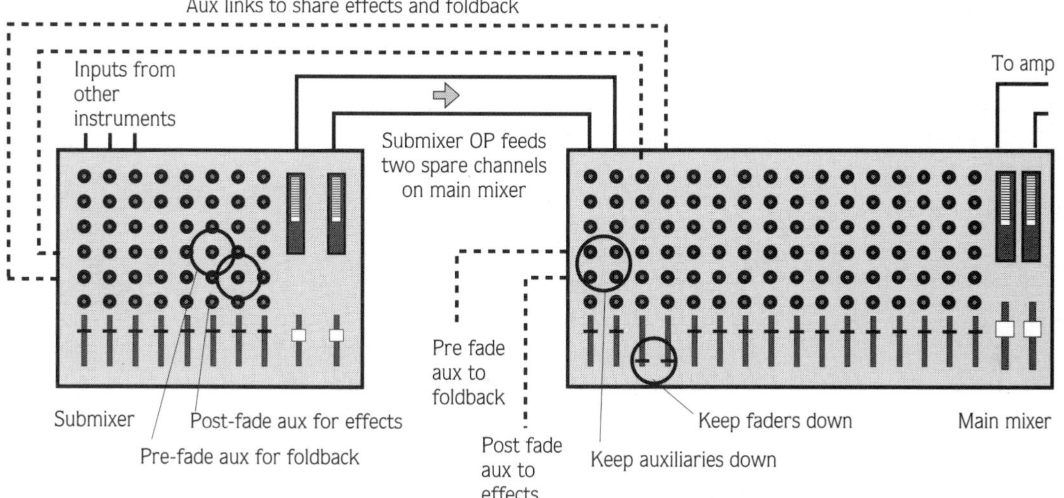

Recording feeds

Recording feeds, are best taken independently of the PA channel fader mix as this often just supplements the on-stage sound. So the on-stage sound may be missing in the recording, leaving you with a very vocal-orientated recording.

Also for recording it is important to deliberately add some audience ambience, otherwise it might as well be a studio recording, and take special care to make it sound like more than the mandatory five people, as so many recordings seem to come out. If you can put the audience feed on a spare track (even a cassette or a DAT to fly in as a last resort) then good, otherwise make sure you monitor away from the direct sound of the gig so that you can balance the crowd noise properly. If in doubt you should err on the side of caution and rely on dubbing on afterwards.

If someone is monitoring the recording feed then in theory a pre fade auxiliary send is best so that separate control is available. If you do go this

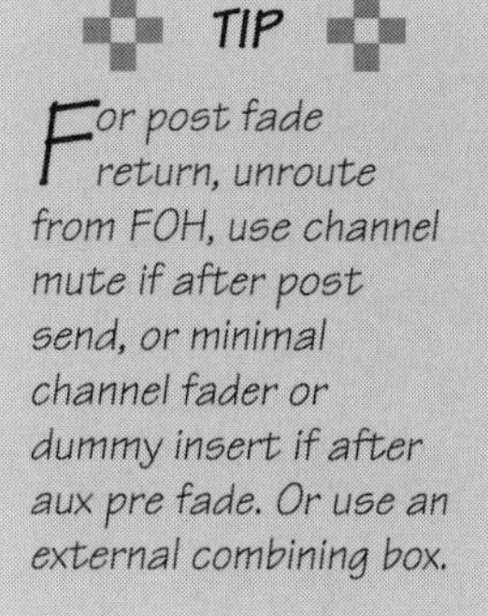

For post fade return, unroute from FOH, use channel mute if after post send, or minimal channel fader or dummy insert if after aux pre fade. Or use an external combining box.

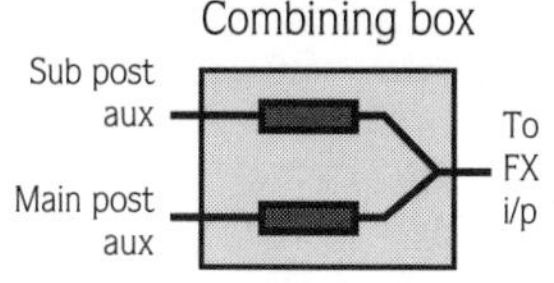

Recording feeds. If separately monitored and controlled, use the pre fade auxiliaries for an independent recording balance. Use two pre fade auxiliaries for stereo. Use an extra channel aux send to include a crowd/ambient mic for the recording – keep it faded down from the FOH mix. For un-monitored recording – use post fade auxiliaries so the PA mix will at least happen. The crowd mic would have to be faded up (for a post fade aux to work) but should be un-routed if possible from the FOH mix.
For multitrack recording use the mixer's direct outs or insert point sends only (i.e. link tip and ring and tap off) or ideally use a separate mixer with extra (or split feed) microphones.

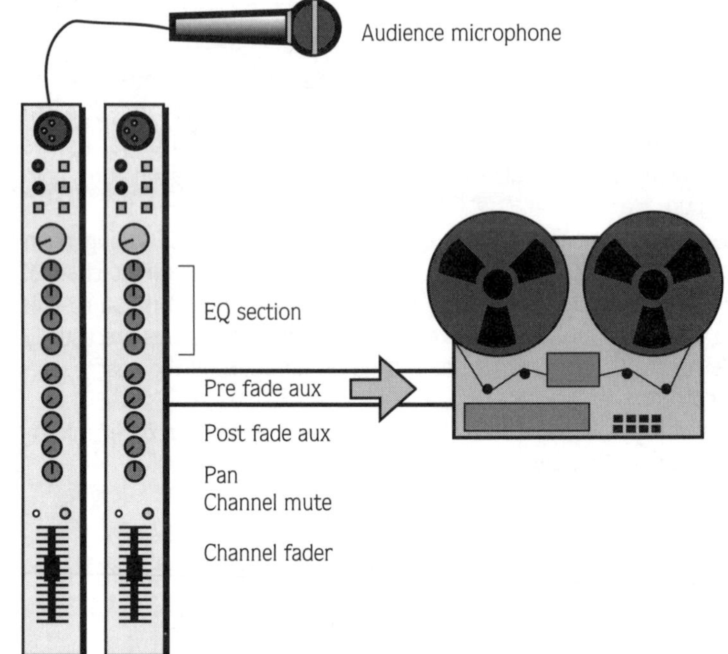

route than watch out for spurious feeds like tape inserts which may be faded out of the main mix, but left running on the channel pre-fade auxiliary. It's a great way to ruin a recording.

However if un-monitored, it could be best to use a post fade auxiliary send so that at least the PA engineer is trying to get the right levels at the right times. You still have the possibility of separate control by adjusting the auxiliary further, and you can certainly set a different balance to the PA sound by initially setting the auxiliaries as required.

Video camera sound

Video camera recordings can also greatly benefit from a mix of the main desk, ideally supplemented with some ambient band and audience mics. Remember the tape replay won't have the benefit of the on-stage live sound which in some cases can be as loud as the PA itself. As well as getting a mixed sound it will avoid the mechanical noise from the videotape, which the internal mics inevitably pick up. If a desk feed isn't available then at least try to rig an external ambient band mic. PZMs are ideal for this sort of thing.

5

The mixer

Intro

The heart of any sound system is the mixer, which is used to route and process the sound. We take a detailed look at the PA and monitor mixers and see how they differ from their recording counterparts.

The mixer

The mixer looks like a complicated beast, but the good news is that it divides into just two main sections – inputs and outputs. The very good news is that the input side divides down in to repetitive identical vertical strips called input channels. Once you understand one of these vertical strips (input channels) you understand 70% of the mixer. So here goes.

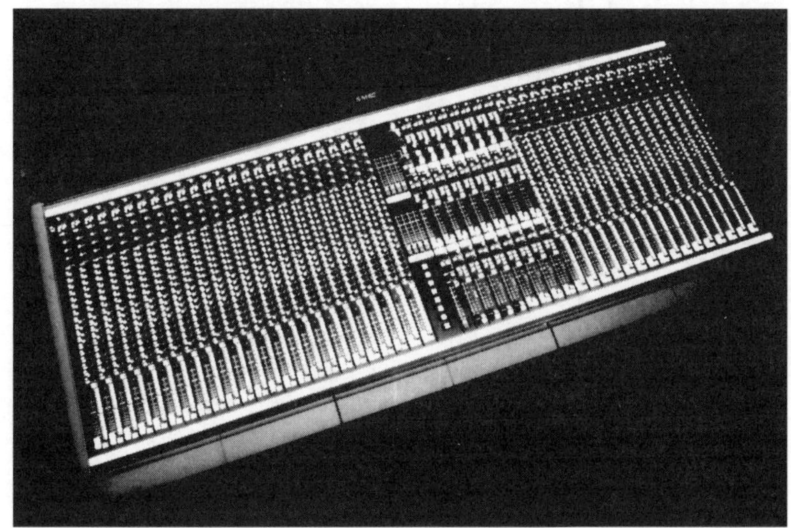

The Soundcraft Vienna II sound reinforcement console

The mixer input channel

As its name suggests, the input channel looks after the inputs, such as microphones and line sources. It matches the source to the mixer and provides level and equalisation control, as well as the routing to outputs (main and foldback) and effects devices.

The signal flow through the channel largely follows the physical layout of the channel. Before we look at this in detail let's take an overview.

A channel needs a means of selecting between different sources and setting the coarse levels of such. An equalisation section to control the frequency response of the source (for corrective and creative reasons), a channel fader to set the level easily, and possibly some routing switches to decide where the signal should go (i.e. groups for different stacks or parts of the venue and for sub grouping).

A typical mixer input channel

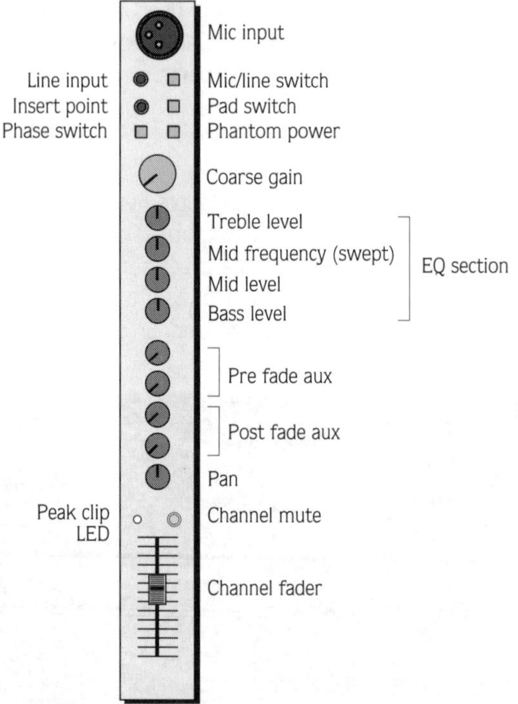

In addition to this path through the mixer, extra feeds are taken off the channel to feed the foldback and parallel effects devices. Each auxiliary is common and shared between all the channels and goes to the same device. It is an easy way of sharing an effect or foldback amplifier/speaker. An insert point is also provided, where external serial effects can be inserted which affect only that one channel.

Serial and parallel effects

A serial effect is one that replaces the original signal with its own effect. Examples of these include equalisation, gating and compression. Some delay based effects such as chorus and flanging may also be added this way if they are only needed on the one source. Otherwise an auxiliary may be used if the effect needs to be shared. Note that in this case the balance control or effect depth control will be used on the unit itself to re-introduce the original if required.

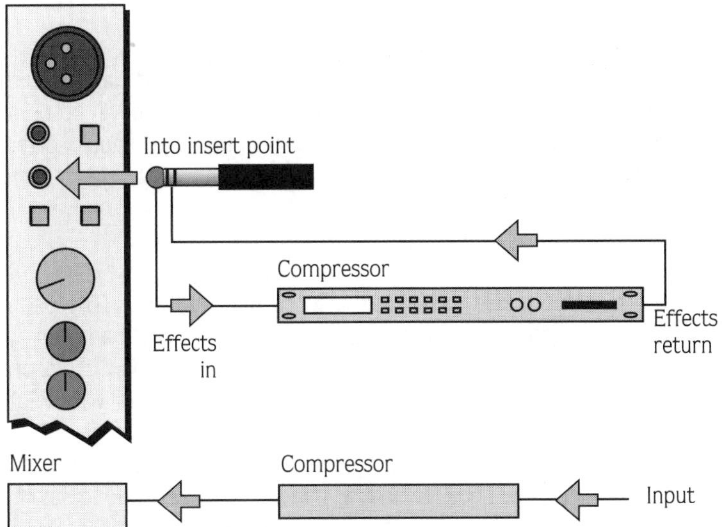

Serial effects (e.g. compression, gating, EQ) replace the original sound with the processed version

A parallel effect is one that is used in conjunction with the original, adding an amount of its effect rather than replacing the original entirely. Parallel effects are most often used on the auxiliary system so that they can be shared between all of the channels. If the effect is required only on one source, then the insert point can be used.

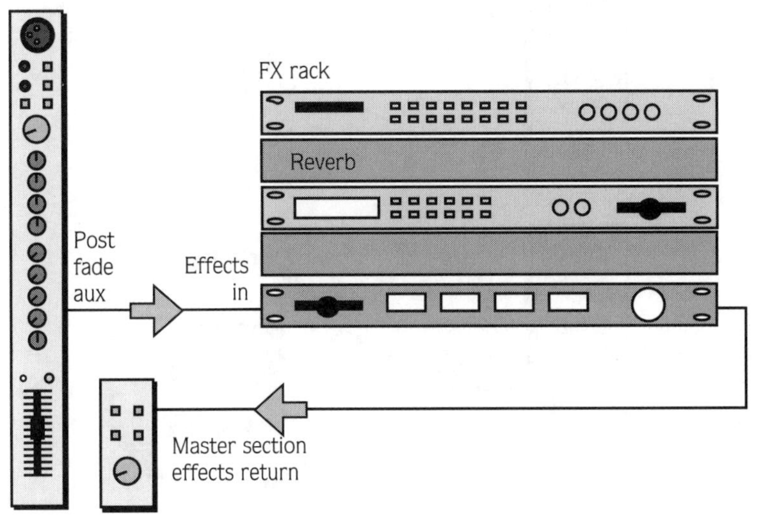

Parallel effects (e.g. reverb and delay) produce effects that add to the original sound

The effects balance needs to be adjusted accordingly. For parallel effects used on the auxiliary system, the balance or effects depth on such units will be set to maximum effect only. Otherwise the amount of the

> **TIP**
>
> *With an unknown mixing console, don't be drastic with the EQ in a live situation. Learn its personality first. If you have to use a mixer you've never used before, don't panic – just follow the channel strip down to find your familiar controls. Be slow and methodical about it. Watch out for tricks like an EQ bypass switch, or auxiliary pre/post switches and group routing switches. Then you should have no problems. Next work out the main output section and find the important things fast – like the master and group levels and the tape repro switch. Take a special look for any disastrous switches like 'solo in place' and group mutes, which could prove to be very embarrassing indeed.*

effect will be reduced or negated. If the effect is being added as a parallel effect but via the insert point as it is only required on one channel, then the effect balance control should be adjusted for the required blend between dry and effect.

Now let's look at the mixer in more detail.

Input selector (mic/line switch)

The mic/line switch determines which input socket and gain circuitry will be used. On a professional mixer, the microphone input will be on a three pin female XLR featuring a low impedance balanced input, often with phantom power available from it. For more about these electronic details see the later chapters.

Pad

Ideally a pad and phase switch will also be featured here. The pad switch puts in an attenuator which reduces the gain by a preset amount (usually 20 – 30 dB) before it reaches any circuitry in the mixer. This prevents overloading the early stages of the electronics, possible with high level microphones such as capacitor types or when miking high SPL instruments such as a kick drum (up to 130 dB). If this still doesn't stop the distortion, you may need to use a pad in the microphone itself (as is sometimes the case with capacitor microphones between the capsule and the mic electronics) or a different microphone capable of these high levels.

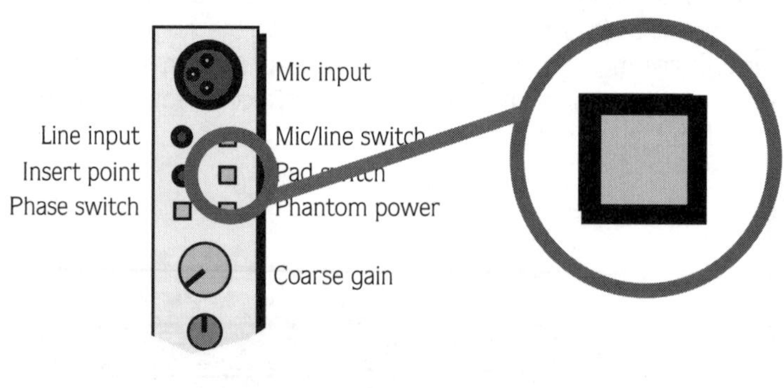

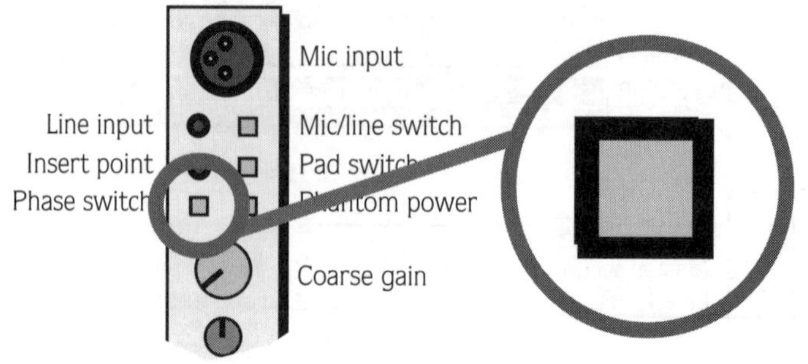

On the mixer input channel, pad (above) and phase (below) normally work on mic inputs only – not line. Note as well as for the stereo mix panning, the panpot also forms part of the group routing (i.e. left to odd (1 and 3) and right to even (2 and 4).

Phase switch

The phase switch is useful for correcting incorrectly wired leads, or when mic placement causes phase reversals, such as over and under miking a snare. In this case the air is being pushed in different directions at the same time and, when combined, the two signals cancel each other out, causing loss of bass (or comb filtering). An adapter lead can easily be made using external cables if no switch is provided, but it is advisable to make it in a special coloured cable to distinguish it. Note this switch often works only on the balanced microphone input and does not affect the line input.

Phantom power

Phantom power is a special consideration. It appears only on the balanced microphone inputs and provides a convenient source of power for capacitor type microphones and active DI boxes. The phantom power is so named because it applies its power (48 V DC) without using extra wires in the cable. For ordinary balanced devices, the voltage has no affect. For instance in dynamic moving coil microphones, the voltage is identical on both sides of the coil (pins 2 and 3) causing no current flow. Phantom devices pick up this voltage from pins 2 and 3 and pin 1 (zero volts earth) and use it as required.

This means it works invisibly. The only time it becomes significant is if using special leads or unbalanced sources, as this could result in damaging them by applying 50 V across them. For electronic sources such as keyboards and guitars, a DI box is used to convert these sources to microphone level and impedance, whilst simultaneously isolating the phantom power voltage. The only precaution is *not* to use custom leads in these inputs, which may short or apply these voltages where they shouldn't.

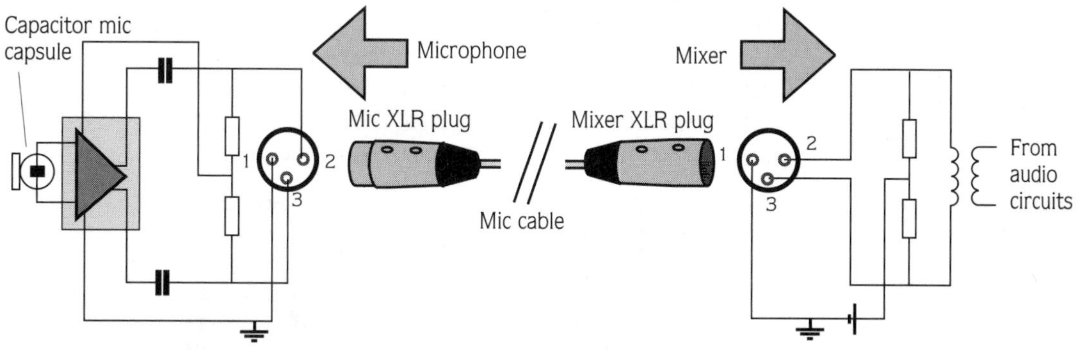

Phantom powering schematic

Gain control

The gain control matches the level of the source to the mixer. You could think of this like the gearbox on a car which adjusts the engine speed to match the road speed. For instance some people sing quietly, others loudly, and line level sources, such as keyboards and guitars, produce much higher electrical signal levels than the microphone.

Gain control

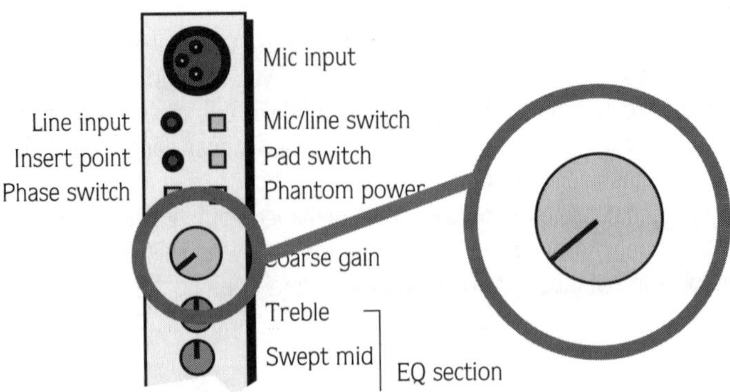

It is possible to get sufficient level out of the mixer without setting the coarse gain properly, but noise or distortion may result (see later section on gain structure). Another factor is that the coarse gain setting can often also act like a tone control in itself depending on how it affects the impedance of the input or feedback loop in the electronics. The correct setting will also help to avoid feedback as it will provide the correct amount of gain excitement for all the following stages.

The coarse gain should be set with the use of the solo button situated near the channel fader. The solo button is also called cue or PFL (pre fade listen) on some mixers. The coarse gain is set so that the solo meter peaks in the red meter section of 0 VU (or 6 on a PPM) during the loud peaks. It should not spend much time in the red. This provides the optimum level through the rest of the channel. The other thing to bear in mind is that any equalisation boosting will also affect the gain and may require reduction of the coarse gain.

Solo button

The solo button is a feed off the mixer channel taken before the fader and usually after the EQ section. This means we can check the level and sound quality independently of the main mix. As well as adjusting the coarse gain in to the channel, it also allows us to check sources in isolation as a quality check, to locate sounds or for cueing before fading up.

Solo button – not available on all mixers

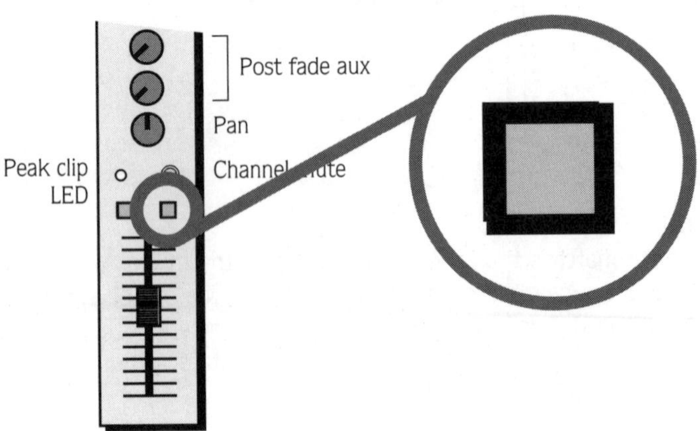

Equalisation

The next stage is the equalisation. This acts much like the tone controls on a regular hifi, but offers more control than the standard fixed frequency bass and treble.

Three-band and swept EQ

Mixer EQ usually has at least three bands – bass, middle and treble. You can think of this as covering separate sections of a keyboard if it helps. With more sophisticated units (sweepable EQ), some or all of these controls also have an associated sweepable frequency control. This allows you to fine tune where you want the EQ to work. For instance do you want the boom or click of the bass drum. Otherwise you have to rely on the factory choice of frequency.

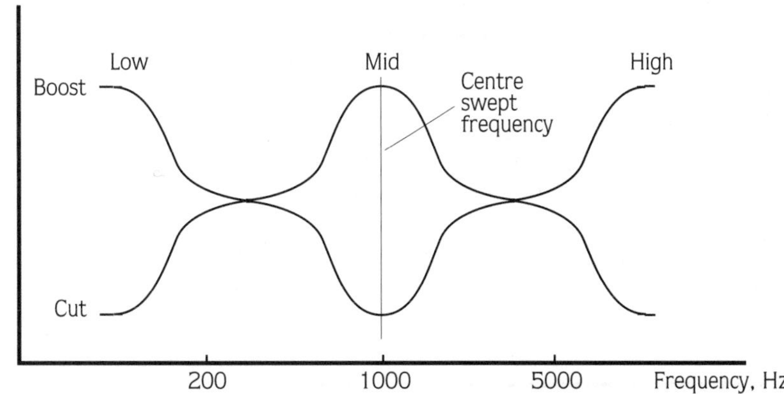

Low, mid and high gain/frequency curves

Swept EQ should be operated much like a radio, putting the volume up (boost or cut – i.e. not 12 o'clock) and then tuning in by ear, the station (frequency) you wish to work on. There are some guidelines as to where the best frequency might be, but largely it should be done by ear.

You may also like to think of EQ in terms of a graphic, which most people seemed to have experienced. Tone controls are the equivalent of a couple of the graphic sliders and that's all you get. With sweepable EQ you still have a limited number of sliders but you can select which position on the graphic you wish it to work. Of course just like a graphic when you adjust one slider, the sliders on either side are also slightly affected. The Q control decides how many of these sliders are affected. The other difference is that EQ usually uses rotary knobs rather than sliders.

The rotary type EQ controls are neutral (have no effect) at the 12 o'clock position. Going past this point clockwise (i.e. 2 o'clock) will provide an EQ boost at the set frequency. Rotating the control anti-clockwise (i.e. 9 o'clock) will provide an EQ cut at the selected frequency.

Equalisation (EQ) curves. With sweepable EQ, the user can adjust the centre frequency. Fully parametric EQ also allows the user to determine the width (called Q) of the frequencies affected

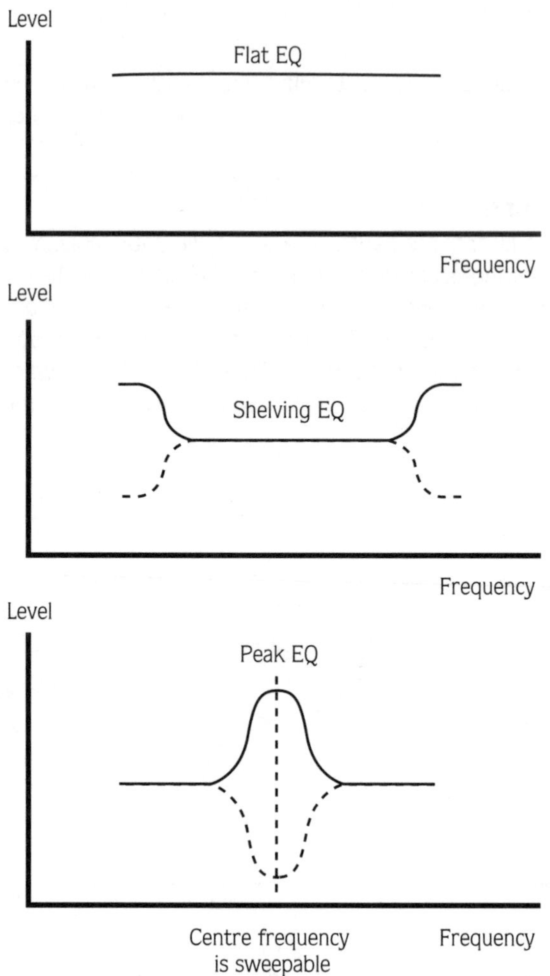

It is recommended you try cutting as well as boosting for a number of reasons. Firstly constant boosting will increase the level through the channel and may cause distortion. Also if the signal is boosted too much it can provide a honk or ringing in the sound. It also tends to increase the chances of feedback, as it adds to the level of excitement.

By cutting EQ you can often get a much more natural affect on the tone of the sound whilst producing a similar result. It also doesn't suffer as much from the problems mentioned above. It can be easier however to find the problem frequency by boosting first and then using the cut position.

Parametric EQ

This is the most sophisticated type of EQ. It has a third control (the Q control) associated with each band, which can be a switch or ideally another rotary control. This controls the range of frequencies around the centre frequency control that will be affected.

It may be helpful if you think of the Q control as the number of notes on a keyboard that are affected by the EQ band. As we said before, the frequency control selects where on the keyboard this starts.

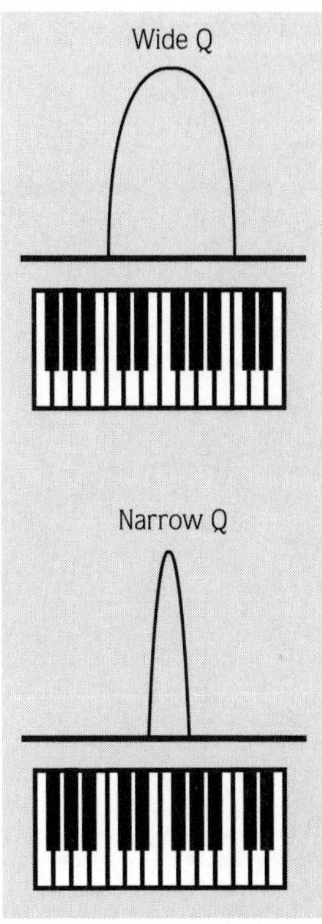

The Q parameters of EQ. Wide Q extends over e.g. one octave, narrow Q over only a few notes

Parametric shelving EQ

Normally parametric/swept EQ has a peak response, but some shelving types do exist. With parametric shelving EQ the sweep parameter decides where the change will start to work from. It starts from the frequency and continues in a shelf like manner. On peak/dip (sometimes called bell type) EQ, the frequency control sets the centre frequency which is affected, this then spreads out in both directions, at a range determined by the Q of the section. On non-swept shelving EQ the starting frequency is factory fixed.

Filters

A filter is another type of useful EQ. A filter works over a range of frequencies starting from the filter frequency and cuts them. Filters come in a range of types:

LPF (low pass filter)

Allows the low frequencies to pass unaffected. The higher frequencies are attenuated.

HPF (high pass filter)

Allows the high frequencies to pass unaffected. The lower frequencies are attenuated.

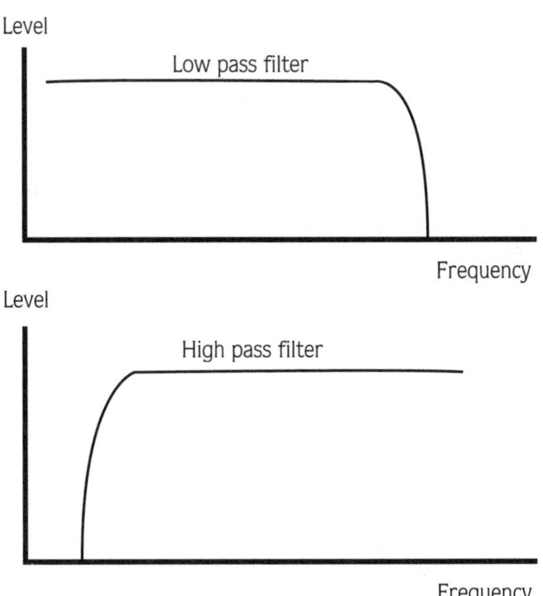

Low pass and high pass filter characteristics

Band pass and band stop filters

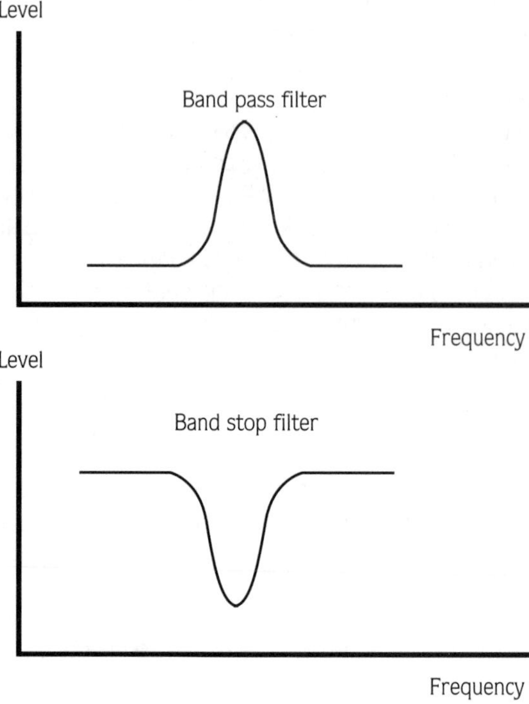

BPF (band pass filter)
Allows only a central range of frequencies to pass and attenuates those above and below the filter range.

BSF (band stop filter)
Attenuates frequencies inside the filter range and allows all others to pass.

Channel fader

After the equalisation the signal is fed to the channel fader. This is a sliding device so that it is easy to operate and gives a visual indication of the level.

The channel fader then feeds the pan pot. This determines how much signal is fed to each half of the pan pot's output. The pan pot often feeds the main stereo output, but, on mixers with groups, routing switches will also be included. These often select pairs of group outputs. In this case the pan pot forms part of the routing, determining whether the signal is fed to the left or right group, equally to both, or anywhere in between.

The channel fader is usually marked with an optimum setting. This position gives some room to increase the level, but mainly to reduce it. This position also sets the optimum electronic level into the mixing bus where all the signal channels are combined. This will give the optimum level between noise and distortion and provides some headroom.

If the coarse gain has been set correctly with the solo button, and the fader is at its ideal position, then the optimum level will be leaving the mixer to the amplifier and the most will be made of the mixer electronics.

Channel fader

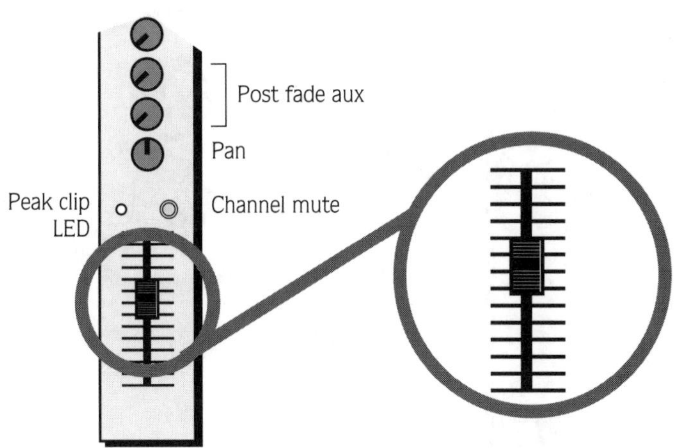

The main stereo outputs then feed the amplifier. If groups are involved then the channel signal feeds the group faders. On PA mixers, these group faders are used to control a selected number of channels in one easy control, rather than having to adjust each individual channel by the same amount, which will take time and may cause inaccuracies. The group may then feed back into the stereo mix and feed the amplifiers, or may be tapped off instead to feed a separate part of the system.

In addition to this path, some extra taps are taken off the input channel. Auxiliary controls take a feed off the channel. They can come before the fader (pre fade) or after the fader (post fade).

Pre-fade sends

Pre-fade sends (before the channel fader) are used for foldback, because their level is independent of the channel fader. This means the foldback mix is independent of the FOH mix. The advantages of this are that the mixes for artist and audience are separate and the engineer doesn't have to worry about one of them affecting the other.

The disadvantages are that the foldback feed will be live even if the channel fader is down, and that any desired changes in both audience and artist feeds must be done on both channel fader and pre-fade auxiliary controls.

Special consideration needs to be taken with this when using radio mics or normal ones which may be switched off or knocked, and equipment which may be unplugged, as they will all be feeding live in to the foldback amps.

Post-fade sends

Post-fade sends (after the channel fader) are used for effects, because the amount of effect will stay proportional to the level. This is the natural method for effects, otherwise as the channel fader is reduced, the amount of effect would stay the same unless both controls were adjusted simultaneously.

Pre fade sends are used for foldback

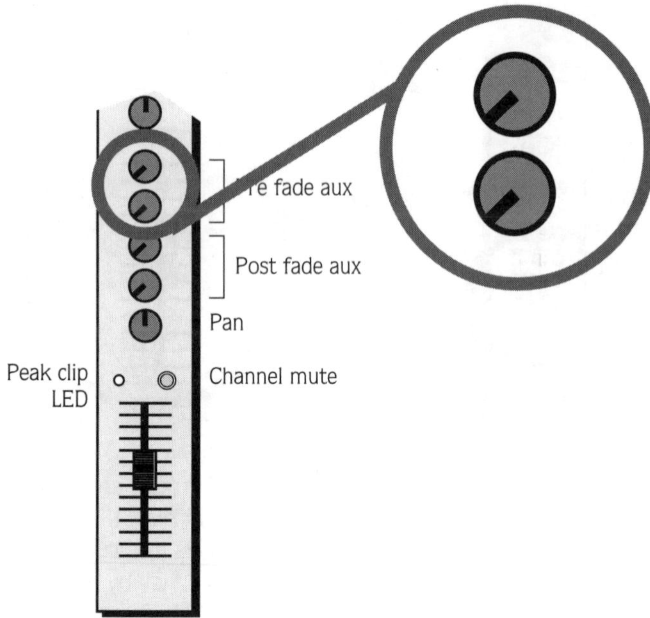

Post fade sends are used for effects

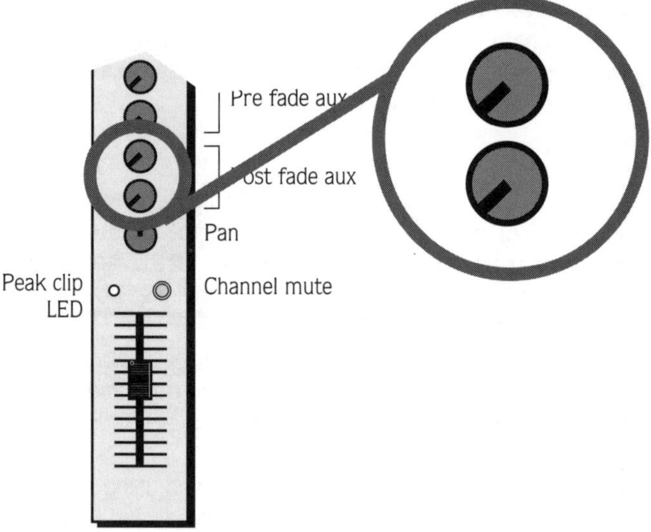

On many mixers, the choice of whether controls are post or pre-fade can be selected with switches, or via internal links inside the mixer. This gives you the choice to decide your own priorities.

Normally the pre fade auxiliaries are taken after the EQ and before the insert points. There are various merits and flaws in any positioning of them. If the pre fade is taken before the channel EQ then it means the artist will not hear the benefits (or drawbacks) of that channel EQ.

The post fade send is normally after the insert points, meaning that the effect send also has the benefit of any external devices patched in.

Routing

The routing options for a PA mixer can vary. The final destination is obviously just a stereo feed to the main PA, and in this case a direct feed to the stereo bus would suffice.

Some mixers offer a button here to double as a channel mute, but care needs to be exercised as only a true channel mute will also cut the auxiliary feeds to the effect and foldback. In such a case using the input selector (mic/line) switch can prove more useful.

Groups

For more complicated applications, you will soon appreciate the facilities of groups. On a recording desk the groups are used as the final level to each track of a tape. On a PA mixer the groups often feed back into the stereo bus directly. On others, like the recording counterpart, a monitor level and pan control is used to direct the group back into the stereo bus. This then also gives the option to pan each group independently of the stereo mix assignment.

In some cases the groups may be used to feed a separate sub system, such as the amplifiers feeding a separate area of the venue, as a glorified volume control. More commonly the groups will be used as sub groups. Any input channel routed to a group will have its final level controlled proportionally by the group fader. In this way sub sections of the mix, almost like mini mixes if you like, can be set up and controlled easily. For instance backing singers or the backline can come up on one (or two for stereo) sub group faders and be controlled in one hit. This saves having to move and re-balance each channel fader individually and is a lot quicker.

Note that although normally located on the right hand side, on some very large mixers, the groups may be placed centrally with input channels on each side. This facilitates access to controls you may use more often.

Group inserts

The other benefit to sub grouping is that if group inserts are featured, it provides an easy way to add an external effect, such as a compressor or equaliser, to that set of sources, rather than having to have a device for each one. This can be very helpful indeed.

Groups as effects sends

Another possible use for groups is as an extra effects send, although this isn't my recommended system as the send level for each channel can't be controlled individually without affecting the main PA balance. However for some spot effects it can be useful.

Groups and aux sends

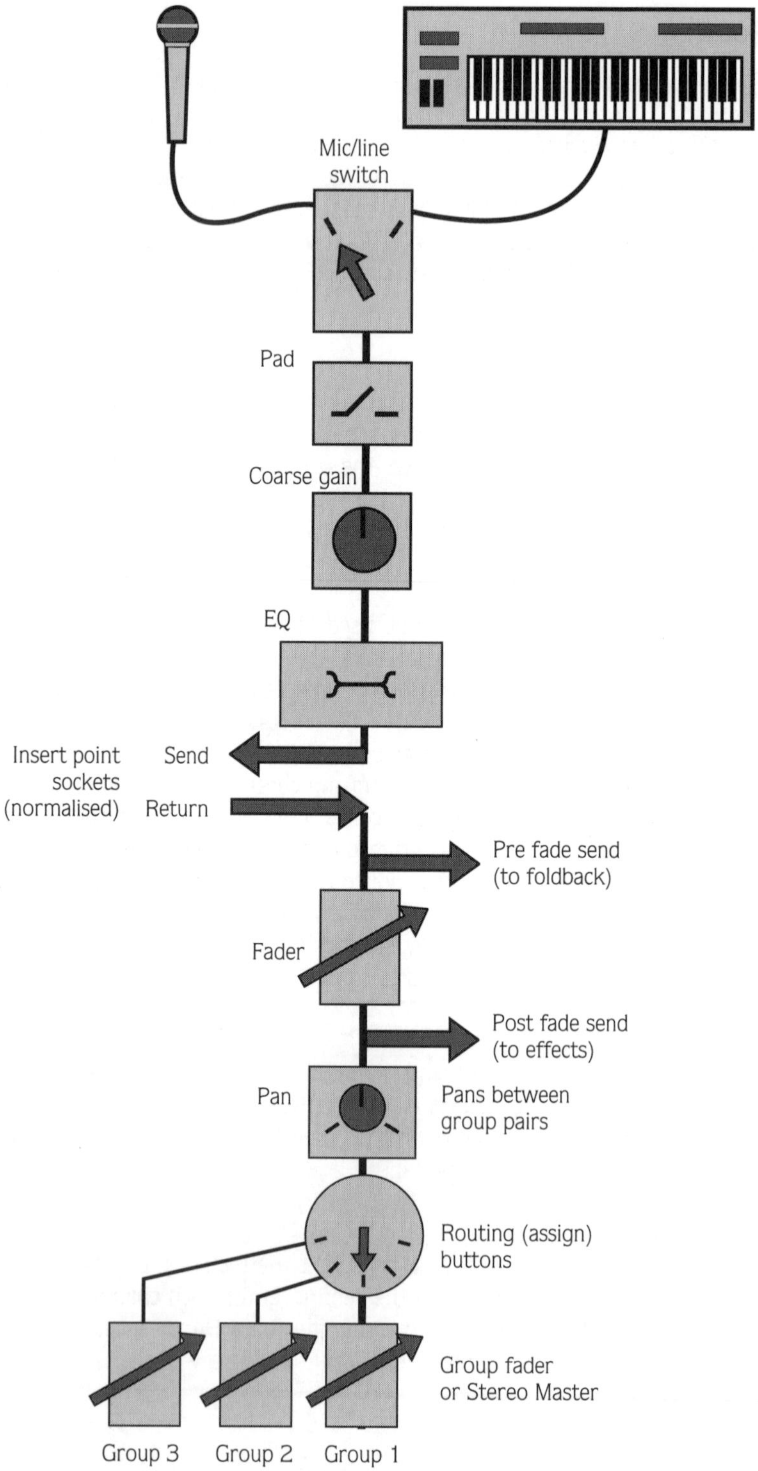

Mic/line switch

Pad

Coarse gain

EQ

Insert point sockets (normalised)

Send

Return

Pre fade send (to foldback)

Fader

Post fade send (to effects)

Pan

Pans between group pairs

Routing (assign) buttons

Group fader or Stereo Master

Group 3

Group 2

Group 1

Gain structure

Most people assume that the mixer's role is just to lump the sound together and look impressive. The mixer actually plays a much more major role than that in the quality of the sound system.

As well as matching the source for optimum gain and noise performance, the gain control can also affect the character of the sound. Many gain controls also affect the impedance of the circuit and cause frequency response changes in this way. Most electronic circuits do not respond linearly with level, and setting the level into each stage again affects its character.

The coarse gain for instance sets the level into the EQ and then to any external devices fed from the auxiliaries, such as effects or the foldback. If not enough level is getting to these devices, the master auxiliary will have to be increased, and will result in more noise in this side chain of equipment.

It is quite easy to get the correct level out of any part of the mixer, as there are so many places it can be compensated for. For instance we could have the coarse gain right up (causing distortion) and then use the channel or group faders to reduce it back to the correct level – but it will still be distorted.

Similarly if the coarse gain is too low, we can use the channel or group fader to boost the level back to normal. This will also mean that the auxiliaries have to be compensated for as well. This results in worse noise performance. The difference between the channel and group faders, or channel or group and master faders can also be mis-aligned. We can have the channel fader too low, causing more noise from the mixer's summing amps, and try to make up for it with the master or group faders. Again the level will be correct but it will be noisier than it needs to be.

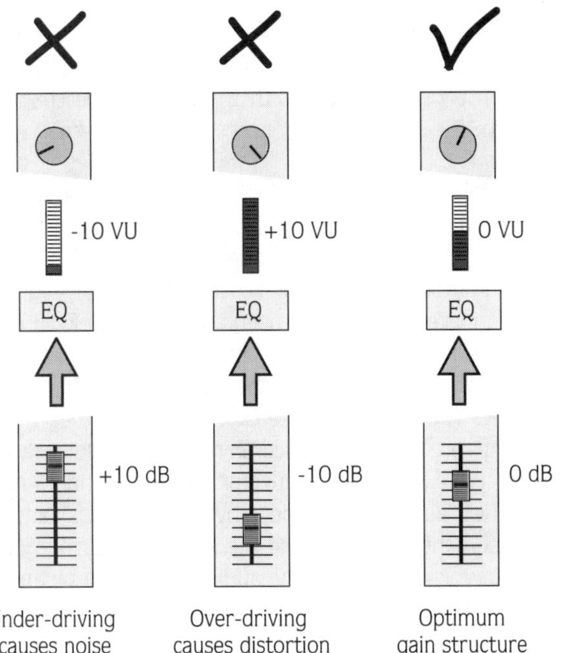

One correct and two incorrect gain structures between coarse gain and fader, all producing the same output level

The mixer normally shows the preferred positions for the faders, and the solo button can be used to set the level in to the mixer. Then it's a case of riding the faders to optimise the levels for the sound you want. For instance some effects units sound better driven hot, others don't. Will you gain from gain structure?

Equalisation swept controls

Equalisation offers perhaps the most radical opportunity to control and refine the sound we hear. With EQ we have the opportunity to almost completely change the sound of an instrument, sometimes even in to that of a different instrument altogether, e.g. fretted bass in to a fretless bass, or an electric guitar in to a nylon strong, or a piano in to a honky tonk.

With fixed EQ the manufacturer has pre-selected the result and we can only decide how much of it we want. With the swept type controls we can tune in to exactly the frequency we want and boost (accentuate) or cut (hide) it.

Swept EQ works like a radio, in that you need to put the volume up before you tune in, else you can't hear where you're going. Cutting EQ can be just as effective as boosting (if not more so) and reduces the chances of feedback. A good technique to find a frequency to cut though, is to boost it first to find it and make it worse, and then to use the cut control.

Recording and PA mixer differences

There are two main differences between PA and recording mixers, namely the type of auxiliaries and the use for sub-groups.

Auxiliaries

A recording mixer will tend to have more post fade auxiliary sends for use as effects sends. By being post fade, it means the level of the effect will go down proportionately with the level of the channel fader – a natural situation for an effect. For recording, external effects are the main requirement.

For PA the main need is for separate pre fade auxiliary sends to use as separate foldback monitors for the artists. By being pre fade, it means that the artist's mix is independent of the main FOH mix. This is the optimum situation as, once the artists are happy, you don't want to adjust things unnecessarily if you need to adjust the FOH mix. This is especially useful once you've set the foldback mix for maximum level before feedback, as otherwise changing the main mix would cause feedback from the foldback. The other main reason for this is that the performers need a different mix to the audience anyway. The audience will want a balanced smooth mix, where as the performers will tend to need to hear more of themselves and the backline (rhythm and bass) and less of the twiddly embellishments. For instance the keyboard harmony may actually distract the main vocalist, but yet the audience want to be amazed at the harmony between them.

The things to remember about this though are that when you change the mix for FOH you may also need to shadow it with the foldback mix – for instance during the lead guitar solo, so that he knows he really is on. Also if you get feedback, say because a performer has moved in front of the speakers, it will usually be the foldback mix which is controlled by the auxiliaries and *not* the main channel faders. Another consideration is that when a performer isn't playing, his foldback mix will still be live, so that any unplugging, or microphone knocking, will come out through the foldback amplifiers. Especially important to remember if cueing tapes fed to the foldback.

Groups

On a recording mixer, the groups are used to control the final level to each track of the multitrack recorder. In some PA installations, some groups might be used to feed different amplifier/speaker sets, in different areas of the venue.

More frequently, the groups will be used instead as sub groups. This means that each channel can be routed to a group. This group then in turn feeds back into the main stereo left and right output to feed the PA amplifiers. Each group allows control of a group of instruments easily – such as backing vocals, backline and keyboards, saving frantic multiple channel riding.

This can also be achieved on a recording mixer by simply setting the tape/group button associated with each group to 'group' and then using the monitor mix level knob to feed into the stereo bus. On a true PA mixer, these switches and level knobs may not be present and will feed the stereo bus automatically.

Monitor mixers

Another type of PA mixer is also available for larger systems – the monitor mixer. It takes a separate feed from the main FOH mixer and its outputs are dedicated to feeding the artist's monitors.

As this mixer is dedicated to the foldback, it mainly comprises a matrix of auxiliaries. Eight auxiliaries are common, allowing for a completely different mix for each performer. The monitor mixer needs to be operated by a dedicated monitor engineer who will need direct visual contact with the artists and won't be concerned with what the audience is hearing. For this reason he may often be at the side of the stage, looking on the performer and watching for visual cues.

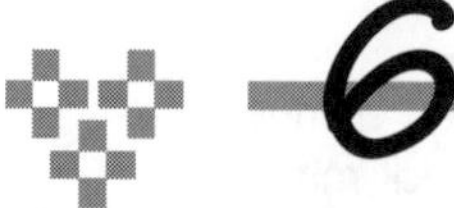

6 Mixing – the art of live sound

Intro

There is a special skill in mixing sound for a live performance and you only get one chance to get it right. We take a brief look at the major considerations and techniques – including sub mixing, grouping and muting, live sound psychology and some tips.

Setting up

Lay out all of the equipment in place. If the flight cases are labelled for their positions it will make placing them much easier, especially if you have untrained help.

Ideally the mixing position will be near the middle of the audience. It certainly shouldn't be at the very back of the hall or under a balcony, as the sound you hear will be very different – each surface will reinforce the bass response by 3 dB, and you will get severe reflections from the surfaces that will change the sound drastically.

Connect all the cables. Start with the power cables first, then the speaker cables, then the microphone cables and finally the processing cables. Try to keep audio cables away from power and lighting cables, or at least make them cross at 90 degrees to minimise interference. High level cables should not be coiled as they form electromagnetic coils and can get hot.

Try to lay the stagebox and any cables up in the air to stop anyone kicking them or tripping over them. Any remaining cables on the floor need to be secured so that people can't trip over them – use gaffa tape or cable mats. *You* are responsible for safety even if you think the cables are obvious. Cables at body height may also need to be flagged with white or fluorescent cloth so that they are visible in the dark.

Turn on the power to everything starting from the peripherals and working inwards to the mixer. The amplifiers should have their volumes set to minimum and should be the *very last* thing to be turned on to avoid any transient thumps.

Play a tape you know through the system and bring up the mixer channel and master levels. Then bring the crossover levels up around half way. Finally bring up the main amplifiers a small amount; make sure they are functioning properly and that all the drivers are working and that no hums, whistles or DC are in the system. Then you can bring the crossovers and amplifiers to a more reasonable working volume.

Check that the microphones work. Having laid out your microphones logically across the mixer, such as drums together and vocals copying the stage left to right layout, you have to set up the mixer. First you should check that each microphone is live by soloing it and listening for ambience noise. If you have a helper he can identify each microphone, but of course you know where they come up don't you ?

Then check that the monitors work, again at low level at first. Then bring them up to a working level. Using the tape is the easiest way to check this.

Next you need to equalise the main system and then the monitor system.

Equalising the system

If the system is bi-amplified or tri-amplified, you first need to set the crossover points and then the volume of each band. This in itself will dictate the broad frequency balance of the system. Whether the system is bi-amplified or not, you need to start with an average level for the venue.

Then you should set the global EQ using a 1/3 octave (31 band) graphic (or the main desk EQ if that's all there is) on the outputs. There are two main requirements here:

1 to achieve a good level before feedback, and
2 to make up for any spectral deficiencies in the system and to get the tonal character that you want. There are a number of ways to do this.

Method 1

This is a very stringent test as the microphones won't be placed as badly as this and can lead to over equalisation, but it is a very quick method.

Place a microphone in the middle of the audience area (at least twice the distance away from the speakers as the speaker's height, so that you get a balanced sound from all the speakers) and then bring up the level until feedback starts to occur. Then you need to find the offending frequency empirically by reducing each graphic band individually. If it is not that band then restore it to flat. Once found, reduce the slider a few extra dB than needed to stop the feedback. Then increase the gain till feedback occurs again.

If the feedback is at the same frequency then reduce the slider more. Repeat until a new feedback frequency occurs. Then locate and reduce this band and repeat the process for this band. You can continue in this way to find more feedback points, but the further you go, the less benefit you will receive. This technique works well for the monitors too.

Method 2
Use one of the vocal mics with the classic 'check 1 2 3' and adjust for maximum crispness and punch. Pay special attention to any ringing or honking and adjust the graphic as necessary.

Method 3
Play a piece of music you know very well and adjust the graphic by ear.

Method 4
The scientific method is to use a spectrum analyser and pink noise generator. Use the microphone as in method 1 and adjust the level to just before feedback. Then using a mute switch, bring in the pink noise generator momentarily and adjust the graphic to reduce the hanging overtones as much as possible. This simulates the real effect of loud music in the environment.

Method 5
Use the microphone to read a steady pink noise signal and adjust the graphic for a flat line. Although this method is scientific under such controlled conditions, it may not prove so useful in anger in the real world. In fact the flat pink noise will tend to offer a subjectively light low end and heavy high end response. It only serves as a guide and needs to be fine tuned by ear. After all, the venue will be full of people (hopefully), not spectrum analysers.

Sound check

To get in favour with the band, you should try to optimise the foldback monitors first

Finally you need to do a sound check with the performers. The sound check should be kept under tight control, and not used as a free for all rehearsal. Once the sound check is finished the band can rehearse as much as they like.

If you do encounter any problem during the sound check, then work around it and come back to it later – you need to keep the flow going or the performers will get bored and start messing about.

If the band is playing you can quickly set the levels using the solo buttons to set the coarse gain and any initial EQ. The concept here is to get something happening as quickly as possible and then to refine it as the band continue to play.

You should start the sound check by testing each instrument individually so that you can EQ them easily and check the quality and level. Then you should check them together, refining the EQ when combined and setting the balance between instruments. When you start to balance, you should concentrate on the loudest required source first (probably the vocals). This means you can match the other levels to it, rather than running out of fading room and trying to push everything up past the end stop.

The drums should follow a close second as this will be the anchor for most things. They are also one of the hardest to balance as there are so many elements. The bass and other backline come next, and then the other sections like keyboards, brass, solos and harmonies.

Sub mixing and grouping

Sub mixing is a useful technique to make mixing easier. By pre-defining some sections and reducing the number of faders, mixing can become much simpler.

This sub mixing may be done externally, like a keyboard sub mixer, or by using the sub groups on the mixer itself.

Muting

Loudness is a subjective comparison between loud and quiet. The ear soon gets used to loud sound and treats it as normal. So for maximum impact you need to keep the lows quiet because the loud can only be so loud before your ears, or the PA, pops.

In addition to the apparent loudness improvements, removing unnecessary sounds makes the mix cleaner and cuts unwanted ambience and spill

Spectral mixing

A good mix can only be defined by ear, but guideline techniques have emerged over the years. Spectral mixing is a concept of allowing a frequency range for each sound to occupy. Rather than letting sounds compete for the same frequency energy, they are interleaved in separate regions by selective use of equalisation.

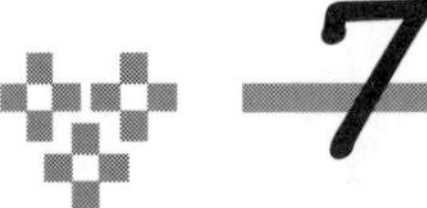

7 Processing

Intro

Fine tuning the sound so that it sounds good at thousands of watts is not a five minute skill. This chapter deals with the multitude of effects processors available to enhance and change the sound – including reverb, delay and delay based effects, pitch shifting, compression, gating, and the most powerful of effects – equalisation.

We can use most effects processors correctively or creatively. In reality we'll probably do both.

Equalisation

Equalisation is the most powerful of effects at our disposal, even though it might not be the most obvious. After all a well EQ'd sound won't be one where you notice the EQ, whereas most people can notice a flanging effect on a sound.

We have already looked at sweepable equalisation and graphic equalisers in a previous chapter. Just remember the tips – cutting can be as effective as boosting and less destructive in terms of feedback. Sweepable EQ acts works like a radio, adjust the volume before you can hear the tuning, and to locate a frequency to cut, try boosting it first to make it easier to find it.

Filters are a special case of EQ and can be very useful in PA terms. By filtering out low sub bass, which the system may not be able to handle, with an HPF, will help to clean up to the sound.

An LPF can be useful for removing some hiss and wind noise from a signal, but excessive use will also reduce the treble in the sound as well. Even bass sounds such as kick drum and bass guitar carry quite a lot of high frequency harmonics in their sound up to 10 – 15 K.

A band pass filter is much like an LPF and HPF back to back. The main application of this is to limit the range say of a vocal, either to simulate a telephone call (300 Hz – 3500 Hz), or in factory PA systems where all the power needs to be directed to the intelligibility rather than the quality.

Band stop filters, sometimes called notch filters, can be very useful for removing a particular single frequency and as they tend to have a very narrow range, won't affect the rest of the sound much. They can be very useful in handling feedback as we'll see later on in Chapter 12.

Compression

Compression is a very important aid to dynamic (level) control. By compressing a sound by even a small amount, we can gain very useful level, which would cost us many pounds in amplifier power. Compressing a sound does a number of things. As well as controlling the dynamic range providing a higher average level and controlling peaks to avoid distortion, it offers subjectively louder sounds, tighter frequency content, a punchier high energy sound and improves dynamic presence in the soundfield.

The Drawmer compressor

A compressor works much like an automatic hand on a fader. When the signal is too loud it turns it down and afterwards returns it to normal. There are a number of controls associated with the compressor.

Attack time

This is a measure of how quickly the compressor can respond to turn the level down. Whilst many people favour the fastest setting of the attack time possible so that no peaks are allowed through, a compressor can be used creatively by slowing down the attack time slightly. This will have the effect of letting the original attack transient through (which most systems can handle) and then controlling the remainder. This gives the sound a lot more punch if the threshold (see later) has been set correctly. Also a fast attack time can sometimes introduce a click or thump at the beginning of the sound, and a slower attack time will reduce this.

For special effects, very slow attack times coupled with a fast release time, can give some interesting amplitude envelope effects, but these tend to be more useful for recording applications than PA.

Release time

This is the time it takes to return to normal gain after the excess sound has gone. The effect of the release time is to reduce the side effects of the compression and make it less noticeable. If the release time is long then the compressor has the best opportunity for consistently shifting the dynamic range and compressing it as much as possible. This can cause noticeable pumping as the level is constantly being adjusted up and down. By extending the release time, once the compressor has reduced the gain, it will keep it there. If this allows some more peaks to pass at the reduced level, then it saves the changes in between and the noticeable pumping. However if the following sound was already quiet, then it stays so and the compressor hasn't smoothed out the levels so much. So the release time is

Compressor envelopes. The attack time is the time it takes to react and reduce the gain; the release time is the time it takes to restore levels to normal.

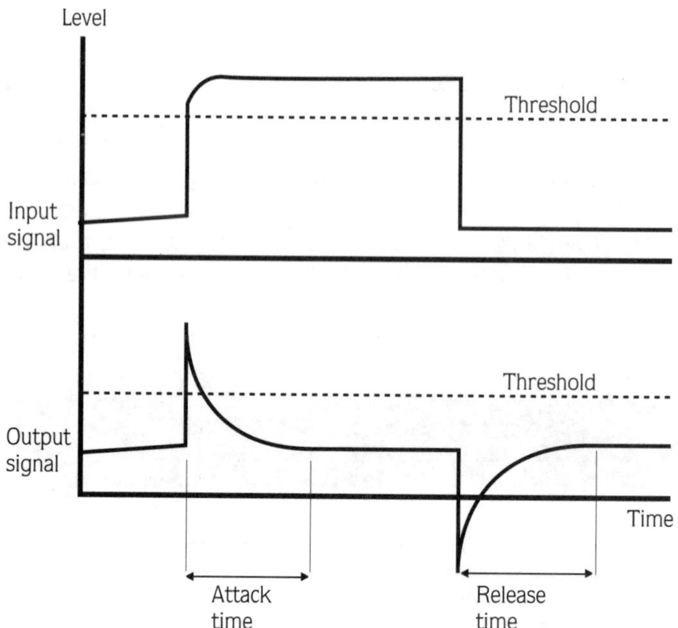

a compromise between short and most effective but most noticeable, and longer and less effective but much gentler. As a very rough guide 500 ms is around a beat (at 120 BPM) and isn't a bad starting point for music. For speech, words tend to be longer, so you can approach the 1 to 2 second mark quite safely

Some units feature automatic settings for attack and or release, by supposedly analysing the content of the input signal and then providing an optimum setting. This can be very convenient and a godsend if you don't have time to set them properly yourself. I prefer to at least have the option for manual control, as any automatic setting removes your control, and your requirements and ideas may be very different to the manufacturer's original concept or musical style.

Threshold

The threshold is the level point at which the compressor will start to control the gain. Below this threshold it doesn't mind what the signal does. By varying the threshold, you can decide whether you want to compress most of the sound all of the time, or just the peaks above a certain level.

Compression ratio

This is the degree by which the compressor turns down the excess. In a limiter design (usually regarded as over 20:1) it will turn it down as much as required so that it never exceeds a certain level – it limits it. This is very useful as an equipment protection mechanism to stop horns blowing up or to prevent overload distortion perhaps to limit the power coming from an amplifier. However it isn't a very artistic option. A smaller compression ratio will help to reduce the dynamic range whilst retaining some of the original dynamic variations, just to a smaller degree of variation.

Compression threshold

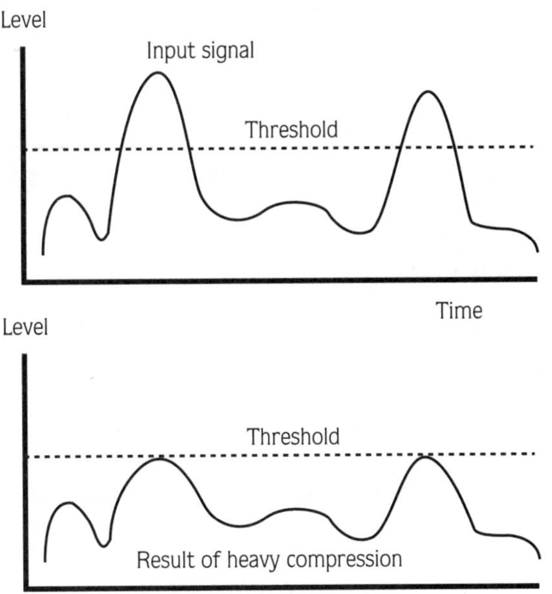

A compression ratio of 2:1 will reduce the level over the threshold by a factor of 2:1. So say we set our threshold for +4 dB. If we had a +10 dB peak, it would reduce the level to 7 dB. With a 20 dB peak, it would reduce the level to +12 dB. You can see how this mild compression can quite drastically reduce the range of dynamics.

For a compression ration of 3:1, our +10 dB peak would be controlled to +6 dB while the 20 dB peak would become about +9 dB.

Compression ratio

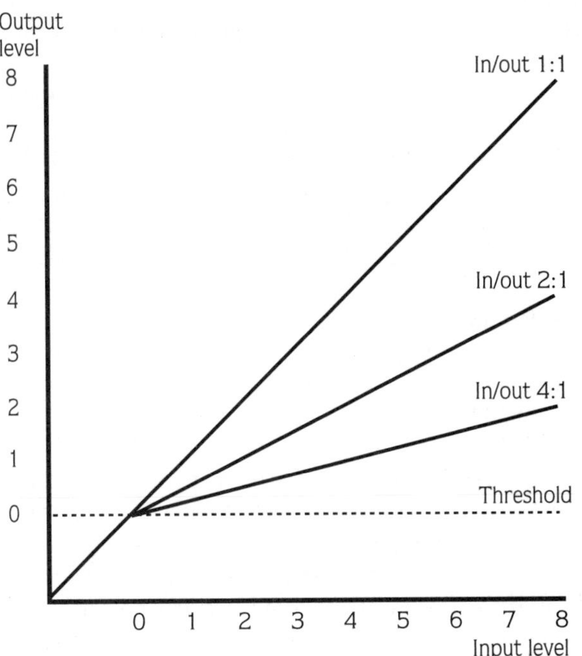

With a limiter set for our +4 dB threshold, our 10 dB peak would become +4 dB and our 20 dB peak +4 dB. As we said, it limits the range to the maximum.

As a guide, the lower the compression ratio, the more we tend to lower the threshold so that more of the range is controlled. With lower compression we want a more gradual change and hence work over a wider range to introduce it.

Makeup gain

The next part of the compressor is a makeup gain control. On some units this may be an automatic part of the threshold and compression controls, but the effect is the same. Because the compression element reduces the level of peaks, it effectively makes the signal quieter. The makeup gain is used to restore the average level. By setting the makeup gain higher, we get an apparent increase in gain, because the signal is amplified on average and only reduced in level during compression. The amount of compression is of course controlled by the threshold control and so we can juggle the results as we desire.

Compressors themselves, even with the makeup gain, are not normally noisy themselves. But as with any device that alters the volume, the noise will also go up and down with the device as it can't distinguish between noise and signal. So in effect adding 20 dB of compression will require 20 dB of makeup gain, which will mean an apparent increase in the noise floor (which it also boosts with the makeup gain) of 20 dB. This effect will be largely masked by the signal itself and it is only during quiet or silent parts that the noise will become apparent.

Having clean sources is obviously beneficial here. One way to improve this is by using a gate or expander gate before the compressor to reduce the noise to nothing when no signal is present (see later). This system can work quite well.

Some gates feature a noise gate in with the compressor. Personally I find these don't work very well as often the gate uses the same VCA element as the compression, which means they are just battling each other. In addition, these gates are usually quite slow, and only the gate threshold control is featured, which means it cannot be optimised manually. I much prefer to use a separate dedicated gate in such cases.

Stereo link

Most compressor units feature two units in one housing. These can work independently or be linked for stereo operation using the link switch. Alternatively single units also have a way of linking by means of a socket on the rear.

The idea is that, if we are compressing a stereo source (using two compressors), the compressor needs to maintain the stereo image. Without linking, if one side were to peak the compressor would reduce the level. The other channel would continue as normal. This would mean that the image would shift to the louder signal and the stereo image would be incorrect. One aspect of stereo works on the principle that the two channels of information contain different levels and the image appears to come more from the loudest side. So if the level changes, we can see that the image will shift. The link switch will bring both sides down equally if there is a peak on either, maintaining the stereo level image.

Side chains

So far we have assumed that the signal controlling the compressor is the same as the input signal, which is usually the case. However, if suitable sockets are provided, we can plug external devices such as equalisation into this controlling chain. It is in fact called the side chain. This allows us to do what is called 'frequency conscious compression'. This is how 'de-essers' work. By making the side chain more sensitive to treble frequencies around the 's' and 't' sounds by boosting the equalisation, the compressor will tend to reduce the level of these through the gate. This can reduce these sibilance effect quite dramatically. As these sounds are only short, it needn't affect the rest of the sound much.

We can of course use an entirely different signal to control the compressor, such as feeding the backline with a vocal feed. Then whenever the vocal comes in, the backline will be compressed and reduced in level. This is how the automatic voice-over function works on DJ consoles. It allows maximum level whilst keeping the voice prominent.

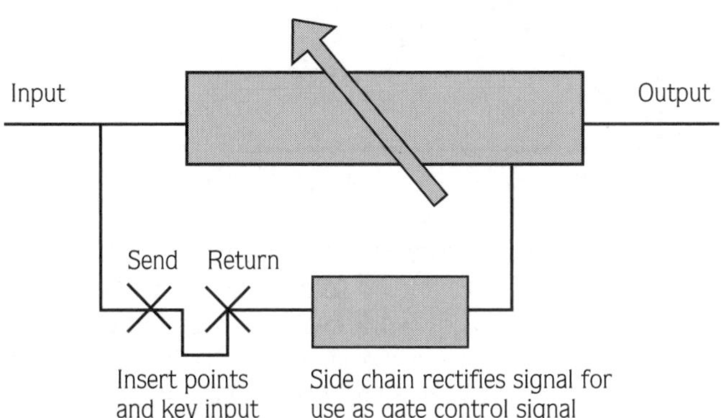

The side chain allows addition of frequency sensitive devices (i.e. EQ boost at problem frequency) for de-essing or external control signals for AGC type DJ voiceover. This is also useful for dipping the backing when the vocals come on in a mix. Note the side chain circuit rectifies the signal, and the degree of rectification can be optimised for peak or RMS response.

Metering

The compressor may include some VU type metering. Often this can be switched between amount of compression or input/output levels. Note that the meter often reads in different directions depending on what it is monitoring. In compression indication, no movement means the compressor is not affecting the signal. If wide movement is seen then you're probably compressing the signal too much. As a guide, around 10 dB of compression will give good control but retain a natural effect. If the effect still sounds too strong, try reducing the compression ratio.

Hard/soft knee

How the compressor kicks in at the threshold point is sometimes selectable with knee switches. With a hard knee the compressor comes in hard and fast. This is required for maximum compression safety such as for limiting.

The soft knee can produce much kinder results as it gradually comes in around the threshold point, often producing a small effect at just below the threshold point. This gradual approach can provide less obvious transitions between compression and no compression. Other compressors say they do this knee selection for you by analysing the signal. It could also be part of the reason why different compressor brands sound so different.

Compression hard/soft knee

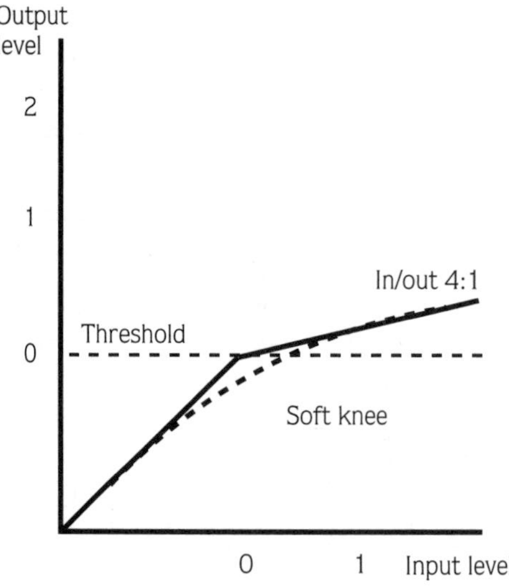

The use of hard or soft knee depends on the instrument, musical style and effect required. As you would expect, for gentle compression such as vocals a soft knee might work best, whereas for percussive sounds like bass and percussion, a hard knee would be more appropriate. Acoustic guitars would tend to need soft knee, while electric guitars depends on the style – perhaps lead, mutes and chops as hard, and strums and picking as soft.

Compressor setup

As a guide, here is how I approach setting up a compressor.

- Set the attack and release controls as short as possible so that you'll hear the effect working easily.
- Decide on a compression philosophy and set the compression ratio accordingly.
- Set the make up gain to an average level and bypass any internal gates.
- Set the threshold to maximum pass through.

- Then feed the input signal in and adjust the threshold to get the amount of compression desired. As a guide 10 dB of compression is a good starting point.
- Then readjust the release time control to get a more pleasing sound. For special effects, lengthen the attack time.
- Refine the compression ratio and threshold to get the effect that you require.
- Finally set the makeup gain to get the maximum level required in to the mixer or amplifier.

Compressor applications

Compressors tend to work best on individual sources because they just work on the loudest signal at the time, be it vocals or drums. But of course your budget may not be so accommodating. A limiter on the output of the electronic crossover in a bi/tri amplified system (or mixer if not) will help to protect your system and keep your maximum level. Compressing each crossover output will deliver the correct level to each band and also provide frequency masking, as each band will tend to mask the other when it is working. It also means a heavy bass transient will not reduce your trebles or mids. The monitor foldback should also be limited, although a ratio nearer compression may help sort out the monitor mix.

Vocals and harmonies are prime candidates for compression (perhaps around 3:1) while percussive instruments such as drums, bass and to an extent guitar, might benefit from a 5:1 treatment. Keyboards are fairly controlled sources, but keyboard players have a lot of opportunities to override your control by holding level back during the sound check or just turning up the sub mix level. Also many synth patches can be quite variable in perceived loudness, if not actual level, and some 8:1 compression can help to control them and keep them audible in the mix.

Compressors and feedback

When using compressors you have to watch that you don't get feedback as your average level will have gone up. Any makeup gain will still be working and you tend to get feedback when the instrument stops playing.

Compression feedback can be reduced only by raising the threshold to stop the compressor hunting so much. Slower attack and release times can also help here. This may mean you need to change the compression ratio as well (raising it) so that the amount of compression stays similar.

If you use subtle amounts of compression depth you will have less problems with feedback.

Gating

A gate is a lot like the inverse of the compressor. It uses the same automatic hand and fader concept, but in this case it's normally closed.

The first thing to understand about a gate, is that it is of use only on sounds below the threshold. It will not remove noise from a source when

Gating. Top picture shows original with noise most noticeable in silences. Lower picture shows reduced noise floor during these silences after gating. Note that it does not improve the noise when the gate is open

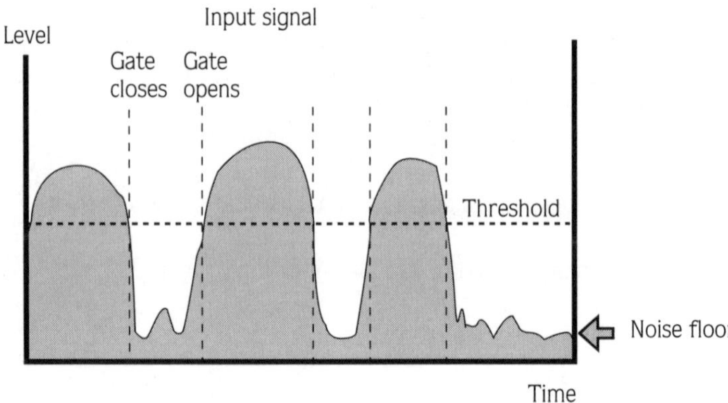

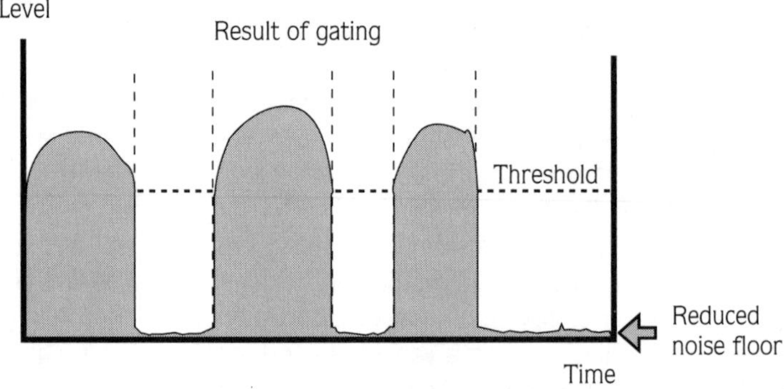

it is playing, only in between the times (when it is not). Gates are not intelligent, they work purely on level. Below the threshold they cut off and take the noise with it. Above the threshold they open and let everything through.

However as we mentioned before, apparent loudness increase can be achieved by reducing the ambient noise level, and for this purpose, and for keeping the mix clean and free of extraneous noise and unused open microphones (i.e. toms, backing singers), they are ideal.

Gate controls

Attack

The attack time is how long the gate will take to open once a signal exceeds the threshold. Often we will have this control as short as possible, but again for creative applications we can use it to deliberately remove transients from sounds (slap bass to fretless).

Hold

On many gates an extra envelope control exists called the hold control. This is an intermediate time when the gate will not be trying to open or close, but just resting. This helps the gate to stop fighting with the signal. Generally it is set to remain open for a little less than the sound it should pass, with the remaining time being faded with the release control.

Release

The release control determines how quickly the gate will close after the signal has dropped below the threshold and will control how quickly the gate will clean up our noise. If too slow it will let noise through, too quick and it may cut the sound unnaturally.

It is quite easy to miss parts of the sound, by incorrectly setting the attack and/or release times, as well as just setting the threshold incorrectly.

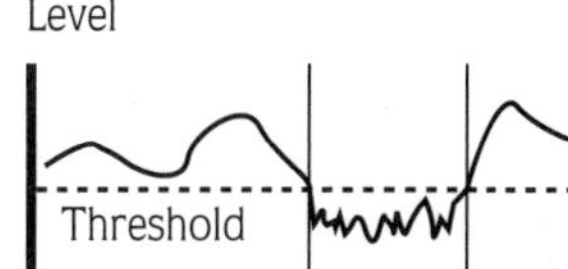

Input signal

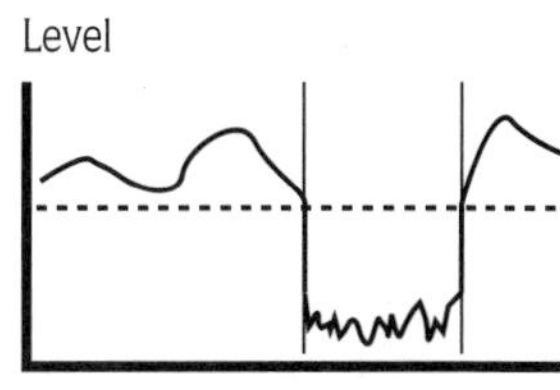

Effect of regular gate

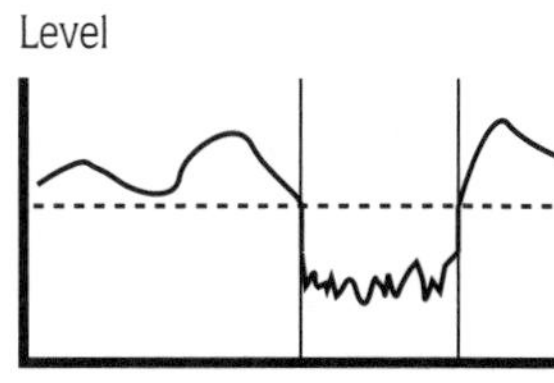

Effect of 1:2 expander gate

Expander gate reduces attenuation by varying degrees, depending on the ratio value

Expansion ratio

On a regular gate, the gate is a switch device, being open or closed. There are no states in between. We can decide how shut the closed position is by setting the attenuation with the depth control. However it only has these two states – open or attenuated.

An expander gate is like the counterpart of limiting to compression. The regular gate is an inverse limiter. The inverse compressor is the expander gate. Instead of being open or attenuated, the expander gate allows transitory depths between these depending on the input level.

So for a gate threshold of – 20 dB and an input signal of – 30 dB, a depth of – 60 dB and an expansion ratio of 1:2, the result will not be – 60 dB as with a regular gate but – 45 dB because of the 1:2 ratio.

This is calculated as – 60 dB – (– 30 dB) = – 30 dB / 2 (as in 1:2 ratio) = – 15 dB. Then – 30 dB + –15 dB = – 45 dB

For – 40 dB in with the same settings as above, the result will not be – 60 dB as with the normal gate, but – 50 dB.

For – 25 dB we would get (60 – 25 = 35/2 = 17.5 + 25) – 42.5 dB.

For +10 dB in, we get +10 dB out, because it is above the threshold. Just checking you're still awake!

As you can see the quieter sounds are allowed to maintain some of their dynamic range, whilst being pushed down lower than they were. This is in fact extending the dynamic range whilst controlling the noise floor. If we assume that the actual noise behind the source is very low (– 60 dB) then we are really talking about expanding the dynamics of our input signal making it more ... well, dynamic.

Threshold

The threshold control determines the level above which the gate will open. Once the level falls below the threshold, the gate closes again.

Depth

For PA use, the depth is a very important control. Because of the transient nature of levels due to performer excitement, change of mic positioning, and even the effect of the audience on the general level of tone, gating can be a nightmare in live situations.

The nightmare can be lessened by going for less ambitious depth control. Changes of 60 dB are not as necessary in live use as perhaps they are in recording, as the dynamic range in this environment is lower, so we can afford to go for less drastic improvements. This means that if the gates fail to work as expected, the sound will not be totally absent, but just attenuated a bit, giving you a chance to adjust it before the audience notices it's not there. The other thing to remember is that a 3 dB improvement is similar to doubling the power of the amplifier, so a gate could turn your 500 W PA into a virtual 1000 W one.

Keying and filters

As with the compressor, we normally have the option of controlling the gate by a signal other than the original source. By equalising the gate's side chain (sometimes called a key input), we can again make it frequency sensitive. This can be very important in PA where the spill level can be much higher, due to the spacing and lack of screens and overdubs compared to the studio.

Often the gate has its own filters built in, or external ones can be used. By filtering the side chain we can ensure it is not triggered by spill from a spurious sound. For example we can stop the spill of the snare keeping the kick drum gate open. Of course it only works if the frequency spectrum (or level) of the spill is different to the trigger signal. With very small differences, you can try boosting the trigger frequency and increasing the threshold instead which may help.

Gate/duck mode

Some gates feature a gate/duck mode switch. The normal setting is gate mode, which is the gate we know and love. The duck mode offers an alternative system, whereby the gate is normally open and closes as soon as it is triggered. This isn't a great deal of use as it is, but if external key sources are used, it can be used like the compressor to dip the input signal, when the control signal plays.

This could be useful for dipping the backing when the vocals come in, or similar. For PA we tend to forgo these features, but if the switch is in the wrong place, your gate won't work as expected, which is why I've mentioned it.

The most important thing to remember about a gate is that it is purely level driven and has no intelligence as to what is noise or signal. With live sound, the levels can change drastically as performers get excited or start to move, so gate thresholds must be watched very closely. Using reduced depths will help to minimise any disasters.

Gate setup

The quick gate setup guide is quite similar to the technique for the compressor, as I hope you can appreciate:

- Set the attack and release controls as fast as possible so that you'll here the effect working easily.
- With an expander gate set the expansion philosophy and set the ratio accordingly.
- Set the depth to maximum so you can hear it working easily and switch to internal key input triggering and bypass or open any internal filters.
- Set the threshold to maximum pass through.
- Then feed the input signal in and adjust the threshold to get the gate open with signal you want. It doesn't matter at the moment if the sound is being chopped off early.

- Then readjust the release and hold time controls to allow the whole signal to come through. For special effects set the envelope controls accordingly.
- Refine the depth control (say to 10 – 20 dB) and threshold to get the effect that you require. It is better to keep the gate open by setting the threshold lower than required although you will usually find the level increases come the performance and the threshold can be lowered.

HINT

Gating the main stereo outputs will have very little effect as something will usually be playing. Gates work best by having lots of them on each individual source. I guess gate manufacturers are just lucky that way.

Creative use of gates

As well as traditional gating to improve noise performance, we can also use the gate for creative uses such as:

- Imposing dynamic envelopes or rhythms on sounds (i.e. chopping vocals with hi-hat aka Kate Bush and Shamen Guitar).
- Auto panning – not incredibly useful in PA as it is better to do it manually with the pan pot if needed. However by feeding the output of one gate into the key input of the other and vice versa, and then having one in gate mode and one in duck mode, triggered panning will occur.
- Dipping – by using external trigger sources and setting the gate depth to a value between 6 and 30 dB we can get a dipping effect rather than a shut off. This can be used to create the rhythmic effects we mentioned above, or to get automatic level dipping in the mix.
- We can also use the key inputs for triggering sounds to make the playing appear tighter. For instance we could gate the bass with the kick drum to ensure they're tight and always play together; bass with sub bass, or gate synth and real brass for similar reasons.
- You can also gate tone or white noise with the kick and snare respectively, to produce larger sounds.

Automatic muting

If you have enough gates, you can use them like automatic channel mutes, helping to clean up your sound and reduce the chances of feedback from unused open mikes. It will take a lot of setting up in terms of level thresholds and it will change through the gig. But it could really help. Again the more subtle you are with it, the less problems you'll get in to.

Reverb

You'd imagine that at a live venue, you wouldn't need any artificial ambience. However PA revolves around trying to avoid reflected sounds which can cause feedback, comb filtering and other interference effects and to get a direct sound from the speakers. Even the close miking effects negate the concept of using the hall's natural reverb. The use of an artificial reverb therefore becomes logical, not only to bring back the natural ambience expected in sound, but also as a small amount of it helps intelligibility and makes it subjectively louder.

We're not used to listening to people in anechoic chambers or fields and are used to hearing the ambience of an environment.

There are of course optimum settings for reverb. Music tends to benefit from 1.5 to 2.5 second decay time, with acoustic music being slightly longer, and speech from 1 to 1.5 seconds. The idea is not to swamp the sound with reverb and to allow it to decay before the next sound comes in. Otherwise we tend to end up with a blanket mush which will tend to smooth over and reduce the clarity.

Brighter reverbs are usually preferable as we have enough low frequency energy kicking about as it is, and the ear is more sensitive to the mid range frequencies anyway. Programmable reverbs will allow you to change all of the parameters, so here's a quick run down.

For PA, use shorter reverb times which allow the music to breathe. You don't want to fill in all the gaps in the music, but shadow the sounds when they stop, creating an ambience. Longer reverb times tend to make the sound linger and hang and can cause feedback.

Reverb

Decay time

This is the time it takes for a reverb to decay by a certain level (– 60 dB). As we've said for live work, music is optimum at 1.5 – 2.5 seconds while speech is best around 1 – 1.5 seconds. In any case don't let the reverb swamp the rhythm of the music. We don't want to fill in all of the gaps with reverb, we just want to shadow the sounds with an ambience.

Be careful with shorter reverb times as they often sound very metallic on some units. In these cases it is better to use the longer reverb times and just use less of it in the mix, or try gating it.

Pre delay

In a natural environment it takes a time for the sound to travel from the source, hit the surface and reflect back. The pre delay simulates this size parameter. We don't often want to simulate a huge environment, but on the other hand pre delay will give a chance for the original sound to breathe before it is shadowed with the reverb.

A predelay of between 30 and 60 ms is a starting point. Longer than that and I would be suspicious, as near 100 ms it becomes a slap back echo in its own right.

HF and LF damping

Any environment will reflect and absorb sound at different frequencies depending upon its composition (i.e. curtains, wood, tiles). The quality of a reverb relies on how the manufacturers interpret these complex reflections with their so called 'reverb algorithm'. Usually the user is given a control to adjust some aspect of this himself, usually in the form of an HF damping control. This basically simulates how bright the room is, or if you like how far the curtains are drawn or not. With PA I tend to favour brighter reverbs so you may want to try raising this figure.

TIP

Reverbs for PA use can benefit from using brighter parameters, possibly with some bass roll off to clean up the bass response which isn't needed so much. Less complex reverbs often work better as the venue will add its own set of complex acoustic paths.

Diffusion

Sometimes we are also lucky enough to have a diffusion parameter. This selects how complex the reverb path is and helps us to create the environment we need. For PA work, I find less diffusion often helps clear the mush, but it is really down to experimentation on the unit itself. The main thing is to be aware that such parameters can make a profound difference and you should take some time to experiment.

Filters

Many reverbs include filters which can be used to reduce the frequency extremes of the reverb. Filtering the bass can help to really clean up the mush in a reverb, although it can leave it sounding thin if used to excess. Filtering the HF (if such a parameter is supplied) is really only useful for creating 'dark' environments more associated with film sound scores than live performances.

Level and balance

I would sincerely hope that the level control will be self explanatory. What might not be so obvious is that is better to set the output level high and control the amount of reverb return from the mixer, as this will give its best noise performance. Similarly the reverb input should be driven hard from the auxiliary sends, as effects are generally noisy devices by nature.

The balance control is used to set the mix between direct and effect. When used with a mixer auxiliary system, we want 100% effect as the balance is controlled from the mixer. If the reverb is used in isolation, such as in–line with an instrument (not advised due to level discrepancies) or on the insert point of a mixer channel or group, then the balance control will need to be used to adjust the effect mixture.

Some units may provide a separate level control or parameter for direct and effect amount rather than balance. The story is the same though.

Delay based effects

There are many delay based effects and most of them can be recreated with any delay line which features variable delay time, feedback and adjustable modulation depth and rate.

For larger venues we need to use delays to time align spaced monitors, but we'll discuss that later.

Here is a quick summary of parameter settings for each effect.

Delay based parameters table

	Delay time	Feedback %	modulation rate	depth
Slapback echo/reverb	80 – 120 ms	60 – 90	0	0
Delay	>120 ms	0	0	0
Echo	>120 ms	50 – 99	0	0
ADT	30 – 60 ms	0	20 – 40	30 – 70
Chorus	30 – 45 ms	0	30 – 70	40 – 80
Flanging	5 – 15 ms	50 – 99	40 – 70	70 – 95
Phasing	1 – 5 ms	40 – 80	10 – 50	60 – 90
Warble	>50 ms	any	5 – 60	80 – 99

Delay effect definitions

Delay A simple repeat of the sound, much like playing it twice. It needs to be long enough to be interpreted as a separate event.
Echo A number of repeats of a sound which decays. Much like shouting from a mountain or canyon.
Slapback echo A very tight echo with few repeats, much like the delay effect between tape heads (remember the WEM Copycat?).
Reverb A simulation of a natural environment such as a hall or room. Only in the open air, such as a field (or an anechoic chamber), do we not hear this ambience.
ADT (artificial double tracking) An electronic simulation much like performing the same thing twice. Because of human nature we never do exactly the same thing twice and subtle pitch and timing variations occur.
Chorus The effect which simulates a number of people playing the same thing (see ADT).
Flanging Like listening to a sound at the end of a very long moving tunnel.
Phasing A sweeter sound than flanging, phasing is a much tighter comb filtering effect affecting the higher frequencies more. Originally it was achieved by delaying one taped version with a hand on the spool. *Itchykoo Park* is the classic recorded example.

Subtlety

No it's not an effect, but if it was you should use it. It means not going over the top, and realising that the sum of the parts is bigger than the whole and that small refinements do make a big difference. If used subtly, any effect, especially flanging and phasing, can produce a very interesting richer sound. If used obviously, they tend to sound like effects (i.e. robots and Martians).

Delay/time chart

If you are going to use a delay or echo effect on a sound, it really helps if it is somewhere near the same tempo as the music. Of course variations can provide interesting polyrhythms, but we have come to desire straight musical delays. You can find the delay by ear, but if you know the tempo of the piece then use this formula:

time per quarter note beat in seconds = 60 / tempo in BPM

Then simply multiply by 1000 for ms markings and multiply by whole numbers to get different note values.

Tempo/BPM table

Tempo (in BPM)	*Crotchet (1/4)*	*Quaver (1/8)*	*Semiquaver (1/16)*	*1/2 4/4 bar*	*1 4/4 bar*	*2 4/4 bar*
130	462	231	115	923	1846	3692
120	500	250	125	1000	2000	4000
110	546	273	136	1092	2184	4368
100	600	300	150	1200	2400	4800
90	667	333	167	1333	2667	5333

(delays in milliseconds (ms). 1 ms = 1/1000 of second)

Warble

Well you never know when you want that underwater or alien voice-over effect do you ?! Essentially very rapid changes in pitch, caused by the modulation rate and depth reading the sound out of the delay at variable speeds. 'Take me to your leader baby !'

Aural exciters

Aural exciters claim to put back what has been taken out and to provide maximum apparent loudness. Although the principle is good I've personally not had a lot of luck using them for PA. By definition they are subtle effects and I find them best used on individual sources when recording.

There are a number of units available and they work in different ways. The original ones tended to add high frequency harmonics, akin to the natural distortion which valves create. Later units then claimed to align the wavefronts and phases of several frequency bands, and others claim a bit of both, adding sub bass harmonics and more besides.

They all seem to do something beyond simple HF boost, but seem quite variable in their success. When overused I feel they actually distract from the sound, and some can produce over pronounced and tiring effects. By their very nature they can emphasise noise, interfere with feedback and contribute to the 'more gear to operate and go wrong' syndrome. If your wallet can afford such subtleties then certainly check them out, in anger if you can. There is certainly no written or brochure specification that can quantify such subjective refinements. Hearing is believing – so use your own ears.

Harmonisers/pitch shifters

Having a harmoniser as part of a PA rig might be going a little too far. They really need to be programmed and operated on a musical level rather than an engineering one, but some background may be useful.

The harmoniser/pitch shifter is a digital device which can automatically create different pitches versions of the original. We used to create this effect using a tape recording and slowing it down and speeding it up. Modern units work in much the same way except that the duration/length of the sound doesn't change. It does this by adding or subtracting bits of the sound. This is quite a difficult process and even the best units suffer from graininess at the extreme ranges. Cheaper units have a job at *not* doing Pinky and Perky impersonations at more than two semitones.

Firstly they are parallel effects, in that we want to hear the original voice plus the harmonies from the unit. We may also want to add this effect to a number of sources, so it is best used on an auxiliary system with its own balance control set to maximum effect.

The original units used a fixed transposition spacing unless altered manually. Unfortunately music is based on scales which don't have fixed intervals for all keys. The latest breeds of harmoniser have scale facilities, where the scale of the song can be programmed in and it applies the necessary transpositions. Some units can even follow the keyboard chords live by using a MIDI connection (see later).

At less than a semitone transposition (less than 30 cents [100 cents per semitone] in fact) the harmoniser can add great body to a vocal and tends to fill in any pitch inaccuracies. The legendary Eventide harmoniser was a standard studio first aid/tool kit for vocals with a setting of 0.99.

Controlling effects

Many modern effects units feature internal memories so that settings can be recalled instantly. Apart from the nightmare of programming them in the first place (or losing your memory contents), this can be a brilliant aid to the sound engineer, especially with a busy set.

Extending this technique further, many units now have MIDI facilities, allowing the memory selection, programming and even real-time control to be performed remotely or in an automated fashion from a MIDI controller or sequencer. We'll look at this more in Chapter 11.

Driving effects

With PA we need to be careful to use every device as best we can. A noisy effects return is amplified and can ruin an otherwise good amplification system. You should drive the inputs to the effect quite hard using a combination of the channel auxiliary sends first, with the master auxiliary send about 12 o'clock.

If the effect has sensitivity switch settings (i.e. + 4/ – 20 dB use +4 dB setting) or different sensitivity input sockets, you should try to use these at minimum sensitivity and use the mixer to drive them. Otherwise we are dropping the high level from the mixer down to the level of the effect, and then using the effect (or mixer itself) to increase the level back in to the mixer. This would result in more noise

All these arrangements can give 0 VU, but only one will sound the best

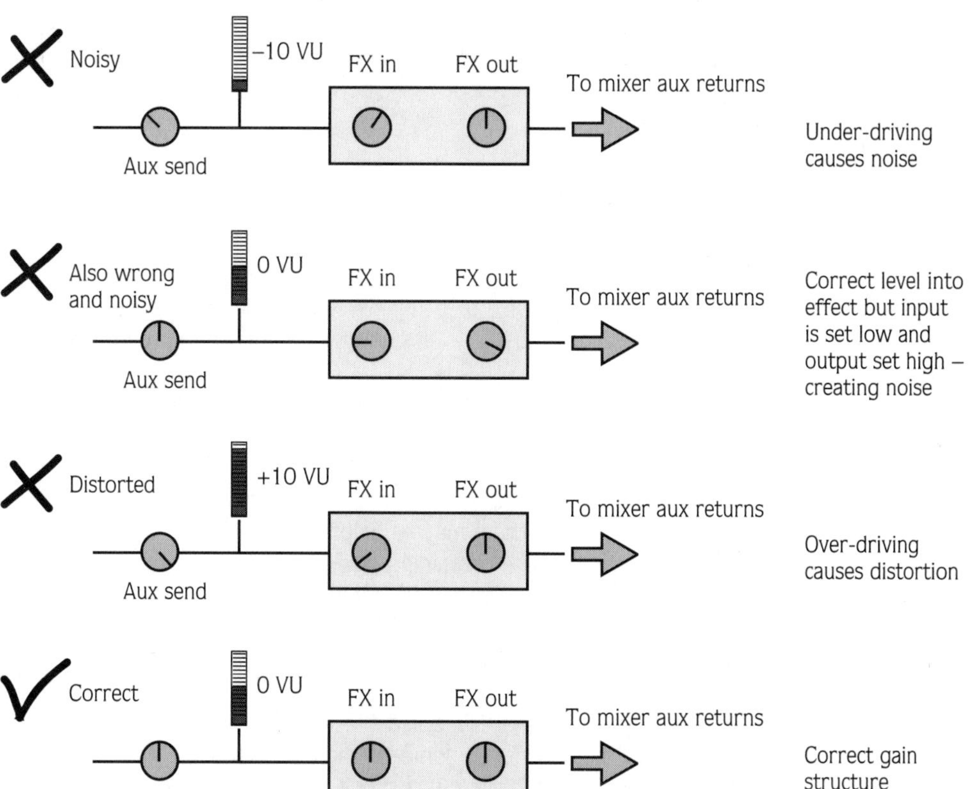

Under-driving causes noise

Correct level into effect but input is set low and output set high – creating noise

Over-driving causes distortion

Correct gain structure

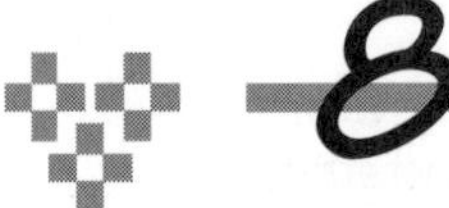

8 Amplifiers and loudspeakers

Intro

In this chapter we attempt to demystify the jungle of amplifier specifications and techniques used – including bi-amplification and electronic crossovers.

An amplifier is pretty much an empty box with a couple of knobs and some meters. The amp is of course key to live sound – otherwise it would all be very quiet.

The actual electronics used in them is of little concern to us, although a general background on them may be helpful. The original valve amp was a nightmare for road use – containing lots of fragile glass tubes that were prone to heat and age – but they sounded nice and warm.

Modern transistor amps bring all the advantages and disadvantages of modern technology with them, but at least they are robust. They tend to sound clean.

A special type of transistor called the MOSFET was introduced a little later. MOSFET amps offered some of the advantages of valve amps but were more robust and fairly low cost. MOSFET amps tend to sound clean and punchy and usually have a cleaner transient response than their transistor counterparts.

The IC amplifier hasn't really been developed yet but no doubt it will come. It has presumably been held back because of the power involved and the heat dissipation needed. Some companies have produced amplifier modules in the order of hundreds of watts with a very good paper specification. Needing only a box, power supply and sockets, it's basically a ready to go amplifier. It's an integrated unit so if it develops a fault, you have to throw it away and start again, but they seem fairly reliable and are moderately cheap. Most have thermal, and short- and-open circuit protection, and they could certainly be considered for cost effective foldback amplifiers or as integrated speaker/amplifier designs.

Power rating

On paper, most amplifiers are very similar in specification. They boast flat frequency responses, low distortion and a power rating which is fairly consistent in its meaning.

The thing to watch with power ratings is the load in ohms at which the rating was measured. Most speakers are 8 ohms, and most amplifier ratings are measured into 4 ohms per channel (two 8 ohm speakers in parallel). This means you'll be running about 70% (multiply by 0.7) of the power in to one pair of 8 ohm speakers, than in to two pairs of parallel speakers providing 4 ohms. Also manufacturers tend to name amps as the addition of both sides in to 4 ohms; so a MOSFET 1000, really means 350 W per side into 8 ohms, which isn't as impressive in the advert of course.

The thing about amplifiers is that they all sound different. Some amplifiers work better with some speakers and some amplifiers just sound better than others with any speakers. The reasons behind this are very complex but undoubtedly have something to do with transient response (sometimes called slew rate), which is a measurement of how quickly the amplifier can respond to the input signal, damping factor (how well the amplifier copes with the back EMF of the speaker and the speaker's changing impedance at different frequencies), and phase response and distortion at all frequencies (in other words how pure it really is). Unfortunately there is little in the way of measurements for these factors and the amp just has to be judged by ear. It is worth doing however, because you really will find that each amplifier has its own sound and character with a certain set of speakers.

Facilities

You don't get much in the way of facilities on an amplifier, and few are exactly essential. Some useful things though are a power light for each fuse of the power supply, a signal present LED, and a signal distortion LED. It would be nice if amplifiers gave some clue as to how hot they thought they were, but this tends not to be a feature.

You can usually hear the fan working so you know when that's come on. In theatre conditions it would be very nice to turn them off, as they do distract from the actors speaking. The change from fan on to off is often more disturbing and it is better to be able to leave it permanently on or off.

Power amp – check where the fuses are (rear panel and internal PCB) and carry some spares!

Fuses

Fuses do blow, usually for a reason but sometimes just because of age, physical shock or because they're cheap. Obviously you need to have a supply of fuses and need to know where to fit them. So it's a good idea to get familiar with the fuse layout and access when you're not in the middle of the gig. It's also worth doing this to check you can actually undo the screws. Some manufacturers seem to employ people who work-out a lot ! Make sure you have any special screwdrivers and don't forget that some amplifiers use the dreaded Allen key, which of course you never have in the right size (Imperial or metric).

Always !

You should always turn off the amplifier and disconnect the mains lead from the unit when changing fuses as you won't know what is live. And you don't want the amp to thump as it goes on anyway. Often there are fuses hidden in the mains inlet and on the circuit boards inside, in addition to the power lead itself, so they're are plenty of places to look.

Cutouts

Some amplifiers use electronic fuses or circuitry that cut out when they get too hot or after a surge. After a while they cut back in automatically and you just have to wait. It shouldn't take very long for them to come back on but it will seem like absolute ages at a gig.

The main reason for amplifiers cutting out is when they get too hot. When driving lower impedance speakers the load is heavier and hence the current and temperature. If you're driving below the rated specification impedance then you've only got yourself to blame. It could also be because of a cabling or speaker fault or an error in the system connection by an inexperienced road crew, or just because you got it wrong when in a hurry. It happens !

Another reason is equipment cooling itself. Amplifiers work best with 1U of rack space between each one and anything else. You can buy blanking panels if you want it to look nice. Amplifier racks with sealed backs are a bad idea anyway because of cable access. Putting a fan in your rack can really help the situation. Also make sure that the amplifiers are away from drapes and curtains which may cover their vents, and don't place them directly in the sun (outdoor gigs) or powerful spotlights.

Monitoring and remote VCA control

It is possible for amplifiers to include self monitoring circuitry which can be connected to a computer system. This is normally coupled with remote VCA control of the volume, allowing the amplifiers to be situated out of the way but still controlled and monitored. Some are even clever enough to detect if the output is distorted compared to the input and alert you to the fact. These facilities only come on the more expensive systems but no doubt they will become more commonplace as prices fall.

Last on – first off

The golden rule of amplifiers is 'last on – first off', much like the headline band. This will protect your speakers from thumps and clicks as things get switched on. It also stops the amp getting any nasty surprises.

After the gig it is good to leave the amp rack open so that it can cool down, or leave your rack fan running, while you pack up all the rest of the gear. This will help prolong its life.

The golden rule of amplifiers is 'last on – first off', much like the headline band. This will protect your speakers from thumps and clicks as things get switched on. It also stops the amp getting any nasty surprises.

Bi-amplification and electronic crossovers

We have touched upon bi- and tri-amplification and electronic crossovers in Chapter 3. The basic principle being that, although amplifiers are capable of reproducing the entire range of frequencies, there are a lot of benefits to be gained from allowing them to concentrate on a particular range. Namely that we can get the correct power to each frequency band, that any clipping (or limiting if used) will be limited to that band and that we can deliver the full concentration of the amplifier to those frequencies resulting in a better transient response and cleaner sound.

Due to the response of human hearing, in theory, compared to the mid range, it requires only half as much of the bass, and the treble needs only half as much again. In practice, the bass and mid need more like equal power and the treble half of that.

We've looked at the crossover frequencies, but we didn't discuss what would happen if we moved them. Well in the bi-amplified system it will change the frequencies going to each amplifier and then in turn to each speaker set. As bass is largely omnidirectional these speakers won't need to be placed that carefully. The mid and treble frequencies however are directional and will have been placed for optimum dispersion without reflection.

If we lower the crossover frequencies we will be asking the bass drivers to try and reproduce these extra frequencies, which they may or may not due with success. But even if they do, these speakers are not placed as they should be for the correct coverage.

If we raise the crossover frequency we will be asking the mid range units to reproduce these lower frequencies and take up useful power from the amplifier driving them. Even if this does work, if the low frequencies now above the crossover point distort, this will produce clipping, which contains extremely high frequency components which may well burn out the speakers.

Loudspeakers

The speaker system is the final link in the chain responsible for delivering the performance to the audience. The loudspeaker is a very innocent looking piece of kit but hides many secrets. As technology goes it's pretty crude, most being some paper glued to a coil of wire, suspended in a magnetic field. Or a close equivalent anyway.

Loudspeaker drive units

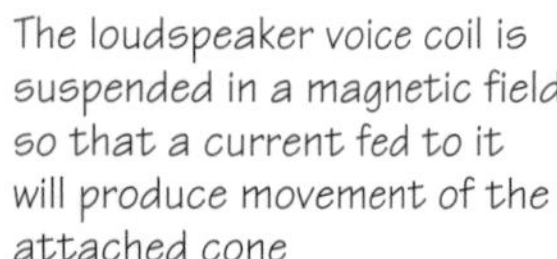

The loudspeaker voice coil is suspended in a magnetic field so that a current fed to it will produce movement of the attached cone

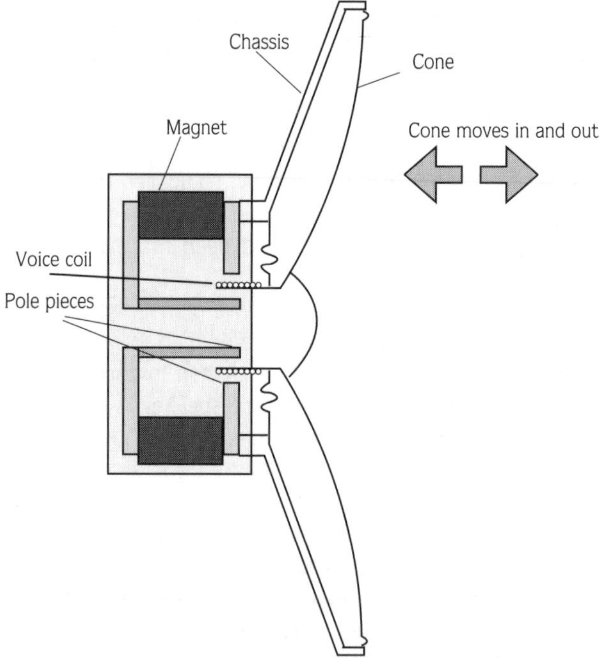

We looked at speakers and enclosures in Chapter 3. Extra considerations are the speaker cable and connections.

Speaker cable should *not* be screened, otherwise the capacitance between the cores will cause serious HF losses and will affect the amplifier and speaker damping. It is also important that the cable can carry sufficient power and that its own resistance is very low in comparison to the 8 ohm speaker, otherwise we will lose significant power. For instance if the speaker cable totalled 8 ohms resistance itself, then we would lose half of the power across the cable before it got to the speaker. This would also make it seriously hot.

It is also very wise to ensure that the connectors used for the speakers are of a unique type and cannot be connected to any other sources accidentally.

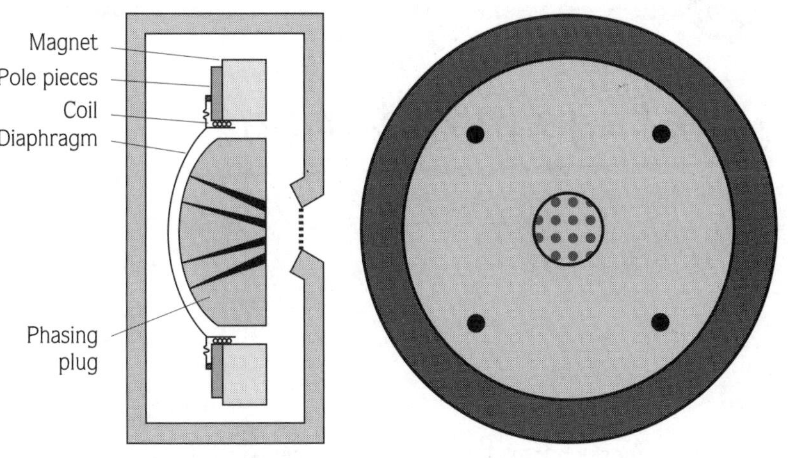

The compression driver handles only high frequencies

Jack connectors are quite commonly used for speakers, but they are not ideal as they don't latch and easily get broken if caught. They are also identical to instrument cables, so increasing the chance of using screened leads and connecting to line equipment by mistake.

Neutrik have recently introduced a plug specifically for speakers (the Speakon connector) which does latch and is unique (see Chapter 17). Failing this a four or five pin XLR plug could be used with a similar philosophy. Any combination of standard 3 pin XLRs as used with the microphone, can lead to incorrect wiring which could cause severe damage. As PAs are often set up in a hurry, the chances of this happening should be avoided. It really is worth the hassle of changing the speaker connectors.

See Chapter 17 for more information on connectors.

9 On-stage monitoring

INFO

Poor foldback can ruin a performance. It can be responsible for poor tuning and vocal pitching and missed rhythmic cues, as the backline can't be heard well.

Intro

Probably the most underestimated aspect of live sound is on stage monitoring and foldback. If the artists can hear themselves properly, then they are much more likely to use a reasonable backline level. This is excellent news as you are then in control of the sound and don't have to compete with the backline. The foldback is also how the artists will judge the job the sound engineer has done – and they pay the bill.

As the monitors can be used very close, we don't need masses of power to drive them. The only problem is that they are very close to the microphones. Luckily they face the rear of the microphone, and with directional mics this means that they will be least sensitive to sounds from this direction. Also if the microphones are used close working, they need less gain in them as well, reducing the chance of feedback.

Out of phase monitors

Another technique is to use two out of phase foldback monitors per position. The microphone is placed between the monitors and due to phase cancellation the two signals from the monitors cancel. The microphone then is less affected by the monitor level. The performer still hears the monitors quite happily because he has two ears quite well spaced apart with his head. Although the monitor sound may be a little wide or floating and may lack some bass due to the phasing, the level improvement before feedback will more than compensate for this.

It is highly recommended that you use a special coloured lead or adapter to do the phase change, rather than dedicate a speaker or fit a switch, to allow for hot backup swaps and prevent accidental switching respectively.

Wedges and sidefills

Monitor speakers are usually wedge shaped so that they can be unobtrusively placed at floor level and aimed at the performer. This keeps the visual line of clutter low and limits acoustic reflections. The speakers should have a fairly narrow field of dispersion. Normal ratings are 100 W to 250 W each, and some models are available with power amplifiers built in to them. This simplifies the installation greatly, although it does little for cross patching if a fault develops.

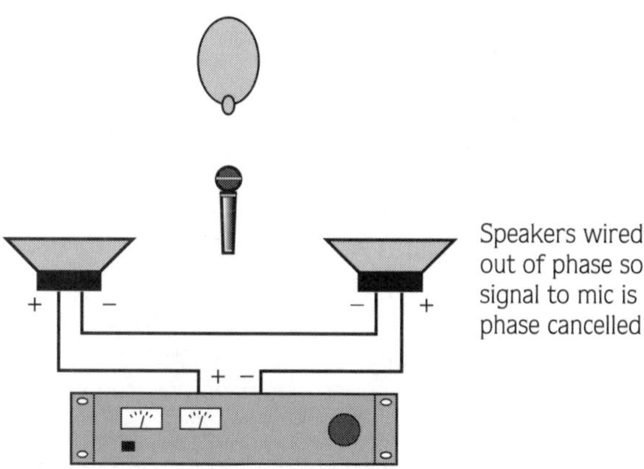

Foldback monitoring. Speakers are wired out of phase to minimise the microphone pick up

The sidefill is another method of getting level on stage although of course it offers a more generic mix for the whole band, so you will still need to supplement it with more personal monitors. By situating conventional speakers at the performers' ear height at the side of the stage, often behind the main speakers, a fairly large amount of volume can be directed at them. If the speakers are the same as the FOH units, then it will help compatibility and EQ'ing requirements and serve as backups.

When using sidefills you may need to angle the microphones away from the nearest sidefill more to reduce feedback. The sidefills will also definitely benefit from a graphic to help prevent feedback. Starting with the graphic set as the FOH unit is a good starting point and then go on to EQ for more level.

Sidefill. Adjust mic/speaker angles to reduce feedback.

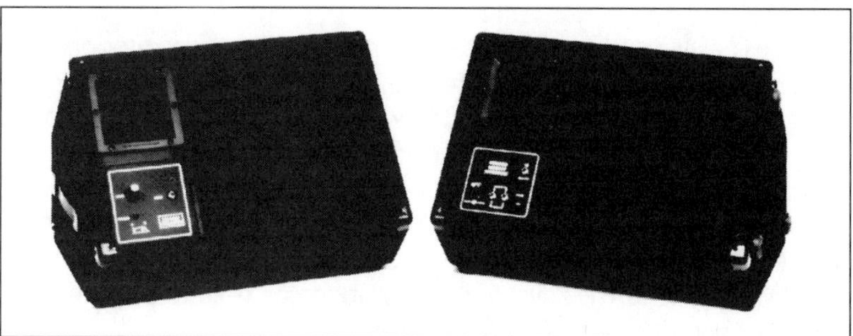

Wedges are typically angled to point at the performer for optimum monitoring

In-ear monitoring

A fairly recent development in foldback has been the wireless in-ear type, such as the Garwood systems. They are quite expensive, but the benefits are huge and the system is already widely used by professionals.

The system uses an unobtrusive ear monitor fed from a radio receiver pack. This gives the performer the freedom to move around from a soundfield and wire less point of view.

By placing the sound in the ear the performer can get exactly the level he needs without any consideration for feedback or spill. It also removes these problems for the sound engineer and requires a lot less gear and cables on-stage. As they provide a quality feed, deafening levels are not required. Care must be taken regarding hearing damage which is still possible with the distances and levels involved here.

The Garwood PRS II in-ear monitoring system gives the individual artists control over their own monitoring levels

Talkback

Talkback is especially useful where a number of sound operators are in different areas (such as a separate monitor and FOH engineer) and of course for theatre sounds. Talkback between the lighting and sound operator and stage management can be helpful for cueing purposes.

There are two types of talkback – open and closed. Open talkback means that all the headset mics are live at the same time. The disadvantage of this is that the ambient noise can be high and anyone talking is heard by the others. The advantage is that it doesn't need to be continually switched and there are no priority issues.

Closed talkback on the other hand requires a press to talk switch on the headset. This reduces background noise and crosstalk, but requires the use of the switch each time.

Wireless versions of talkback systems are available and again they must be DTI approved. The are prone to dead spots, spurious pickup and continual battery replacement. It is unlikely that many operators will need to be so mobile as to benefit much from not having a cable.

It is very useful to have a foldback and talkback feed to the dressing rooms so that the performers know what is going on. A switched talkback from there could also be employed.

10 Playback systems

Tape machines of course have the live impact and stage presence of a brick wall and are not very visually appealing either. The audience tends not to leave the venue saying 'did you hear that solo the tape machine played ? Wow !'

Intro

It is often necessary for a live performance to include pre-recorded elements, e.g. backing tracks, spot effects and inserts. We take a closer look at the benefits of the different systems available – including analogue and digital tape, MIDI sequencer based, tapeless systems and backing it all up.

There are a number of reasons for using prerecorded inserts in a performance. For the performers, inserts offer a reliable medium of support and reduce the need for people and equipment and their associated travel and entertainment costs. It can make the difference between making money doing a gig and not even breaking even.

For theatrical and presentation purposes it provides a tested and optimised sound support without needing to arrange live performers to play on cue.

The best reason is that some things are just impracticable to recreate live, such as fitting a string orchestra in a small venue, doing your own harmonies or overdubs, for adding ambient sounds, or for use during a difficult set of stage movements (i.e. singing whilst dancing).

The Revox B77 – another industry standard machine. Open reel systems are preferred to a cassette system for playback

Tape also offers a very convenient alternative to the time consuming process of having to load long samples in a sampling keyboard, to provide such things as ambient sound effect soundscapes and atmosphere, spot effects (like explosions or media speech), or perhaps even canned applause to encourage the audience.

Any insert can be optimised for sound quality using all of the facilities of a studio, controlled acoustics, overdubbing and time to mix it properly. Also there are no worries about feedback, spill or variable mic positions with a tape.

A disadvantage of taped inserts is that they offer a rigid version which cannot respond to tempo or jamming changes, such as spur of the moment repeats, or indeed requests. The largest problem is that of synchronisation – starting together and playing at the same tempo. The tape really needs to be the master for anything other than short freestyle applications. This involves getting the band to follow the tape, usually by getting the drummer to follow it on headphones, if the tape is not used all the way through the song.

Although in theory it should be easy for the band to follow the tape, it removes the freedom for tempo changes, tempo pushes, and variable length sections. So this method will not suit all performers.

Tape inserts

Tape has been the traditional method for PA playback. Open reel formats are preferred because they can be cued with great precision and offer far superior quality over cassette. People seem to have a phobia over lacing the tape, but it becomes very easy with a little practice.

Unfortunately many standards have evolved for open reel formats, in terms of both speed of playback and the track layout over the width of the tape. The quarter inch width tape has become the normal option, although in studios a 1/2 inch version is used fairly commonly. Professional standards really dictate using only half track tapes, which use the whole tape for a stereo recording (each channel occupies one half of the tape) as it is impossible to edit both sides of a 1/4 track tape for simultaneous use. The domestic standard uses a quarter track system, where the tape can be turned over and played in stereo in that direction too. This is again different from the four track multitrack format useful for quadraphonic and surround sound work

Open reel tape still offers the fastest and most cost effective solution for PA inserts. The ease and speed of editing and cueing. Being able to insert coloured leader tape bands between tracks also helps fast visual cueing in low light conditions.

Open reel tape still offers the fastest and most cost effective solution for PA inserts. The ease and speed of editing and cueing. Being able to insert coloured leader tape bands between tracks also helps fast visual cueing in low light conditions.

Tapes are also less prone to mechanical vibrations such as with records, or of insane behaviour and skipping, as with CDs.

1/4 inch tape format

Name	Common speeds	Comments
1/2 track	15 and 7.5 ips (38, 19 cm/s)	Stereo over whole tape – can't use second side
1/4 track	7.5 and 3.75 ips (19, 9.5 cm/s)	Stereo over each tape side – turn tape over for second side
4 track multitrack	7.5 and 15 ips (19, 38 cm/s)	Quadraphonic/4 channel across whole tape – can't turn tape over (some tracks will play backwards if done – a useful effect ?)

Tape track layout

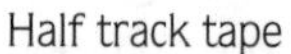

Guard track

Quarter track tape

Track 1 ⇨
Track 4 ⇦
Track 2 ⇨
Track 3 ⇦

Four track tape

Track 1 ⇨
Track 2 ⇨
Track 3 ⇨
Track 4 ⇨

Info

It can be seen that using a 1/4 track tape on a 1/2 track machine will cause the second side to be picked up and played too (albeit backwards).

You will also notice that the common speed between them is 7.5ips (19 cm/s), so using that speed and not recording on the reverse will offer most compatibility. It may still be necessary to have previously erased any tapes using a bulk eraser, or erase using the 1/2 track machine to ensure proper playback on both 1/4 and 1/2 track machines

Using a 1/2 track tape on a 1/4 track machine will pick up in mono on track 1 fairly well. The guard track (spacing between each track) is not always ideal for track 2 playback, so it usually best just to use track 1.

Tape counters

You can *not* rely on any tape counter for cueing unless it is actually reading SMPTE or some other code track off the tape directly. The tape will simply slip against the counter wheel or capstan tachometer and will slip with use – 3 seconds over 3 minutes being typical even on high specification (read 'expensive') machines.

Tape editing

Tape editing is physically a very easy process. It is only deciding where to cut the tape that requires practice. To edit you will need to locate the playback head. This is the head to the extreme right of the tape path when looking at the front of the heads.

The tape is played and paused at the rough location of the edit in point. The tape is then rocked backwards and forwards at a steady speed, in ever decreasing swings, until the exact point is found. This is then marked with a line at the playback head using a chinagraph pencil (yellow or white grease shows up best) which is soft enough not to damage the head as a traditional pencil would.

The 'out' point edit (where you want to join your 'in' point to) is then found in a similar manner. The tape is then pulled away from the heads and placed in an editing block, which is a precision machined block to secure the tape (like a wood mitre). Each mark is then sliced at an angle

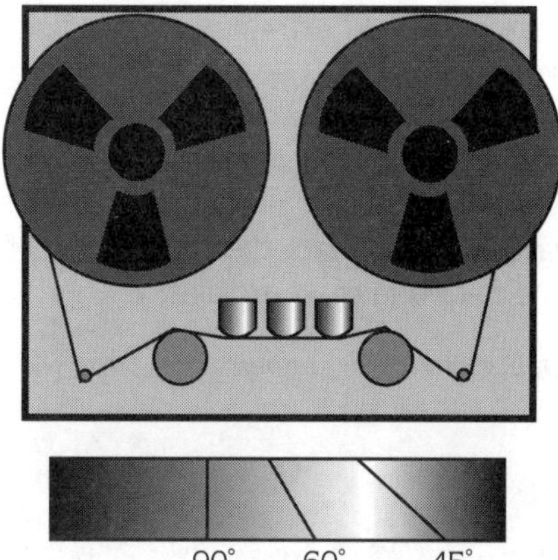

A typical editing block. Mark playback head directly with a chinagraph or pre mark block to a preset guide

(not chopped) with a non-magnetic editing blade (otherwise a click will be heard on replay), and then the two ends are joined using special adhesive tape called splicing tape. As well as being ultra thin so that it will not affect the playback speed as it goes through the capstan and pinch wheel (the rubber and metal wheel that drags the tape past the head at a steady speed), it will not ooze glue or fall apart with age. Don't try to use cellulose tape, it just isn't worth it.

Editing angles

There are often three angled cuts on the editing block. The 90 degree (vertical) one is for butting leader tape, the 45 degree is for smooth mono edits, while the 60 degree cut is our normal choice for smooth stereo editing.

The idea behind the angle is that it provides a smooth transition between the two joins, causing a physical crossfade. The high angle is needed to prevent any noticeable stereo image shifts.

Watch out for the VU meters on the tape machine as they can happily magnetise the razor blades if placed over them, which will then cause a click on every edit.

Tips

It is important to handle the recording and splicing tape as *little* as possible with your fingers, as natural skin oil can adversely affect it. It is advisable to cut off several lengths of splicing tape and rest one end of it somewhere on the tape machine using the razor blade. Then use the razor blade to pick it up again by the other edge. This will also help you to angle it so that it doesn't overlap the blocks channel when you apply it to the tape join. After applying it, smooth it down in the block, gently peel the tape out of the block towards you, and then smooth any remaining air our of the splice by running your finger over it on a flat part of the block.

Editing information

The trick when trying to locate a particular part of the song is that the rocking needs to be quite fast to get sufficient volume and anything like the original sound. The speed of the tape affects the playback pitch. Playing the tape at half the speed produces sound an octave lower, while playing it twice as fast produces sound the octave above. With tape rocking we can achieve wild extremes of speed/pitch, and experience of how the tape sounds at different speeds can really speed up the editing process.

For instance the kick drum, which is normally a well defined click, will be more of a low frequency blur, when rocked slowly. However it is still a distinctive sound and is one of the easiest sounds to locate in editing. All you have to do is locate the correct one of the four (for 4/4 music) in a bar. The following chart explains:

Spacing of kick drum beats on tape (4/4 song)

Tape speed	*38 cm/s (15 ips)*				*19 cm/s (7.5 ips)*			
	Bar (4/4)		*Beat (4/4)*		*Bar (4/4)*		*Beat (4/4)*	
	Time	*Tape Length*	*Time*	*Tape Length*	*Time*	*Tape Length*	*Time*	*Tape Length*
BPM	*ms*	*cm*	*ms*	*cm*	*ms*	*cm*	*ms*	*cm*
130	*1846*	*70.15*	*461.54*	*17.54*	*1846*	*35.08*	*461.54*	*8.77 (3.45 in)*
120	*2000*	*76.00*	*500*	*19.00*	*2000*	*38.00*	*500*	*9.5 (3.75 in)*
110	*2180*	*82.84*	*545*	*20.71*	*2180*	*41.42*	*545*	*10.35*
100	*2400*	*91.20*	*600*	*22.80*	*2400*	*45.60*	*600*	*11.40*
90	*2667*	*101.35*	*667*	*25.35*	*2667*	*50.68*	*667*	*12.68*
80	*3000*	*114.00*	*750*	*28.50*	*3000*	*57.00*	*750*	*14.25*

1 ms of time takes 0.038cm (0.38 mm) of tape at 38 cm/s

1 mm length of tape lasts just 0.2634 ms (263 µs) at 38 cm/s

Note at 19 cm/s you have to halve the length of tape to play with

(Imperial thinkers should divide centimetre values by 2.54 to convert to inches)

Accuracy

With editing it is possible to get an accuracy of around 1.6 mm (1/16 of an inch) so this should be no problem. This is an accuracy of 0.42 ms (at 38 cm/s), which is very high compared to frame accurate SMPTE frame editing offering 40 ms accuracy at 25 frames per second. So you can see that any digital system has to offer sub-frame accuracy which is 1/100th (sometimes 1/80th) of a frame to be useful.

TIP

If cassettes must be used, a chrome cassette is mandatory as these offer a great sonic improvement over ferric types, and reduced head wear over the much more expensive metal types. You may find greater success by having each insert on a separate tape, as cassettes are also notorious for low winding speeds and search accuracy.

Cassettes

In my opinion cassettes are just not acceptable for live use. They cannot be cued and played in with precision, they are generally low quality, and there are many compatibility problems with speed, Dolby noise reduction and head (azimuth) alignment between machines.

Digital tape

Digital tape is generally cassette based and it would therefore pose a cueing problem except for the fact that it offers digital cueing techniques, such as frame accurate time counters and memory locators by reading an extra track on the tape with sub code or SMPTE time code on.

Editing of digital tape requires the use of electronic means rather than physical ones to dub (copy) the signal from one tape to a new one. Although digital dub editing (copying the composite tapes to a fresh tape) does not introduce any quality loss (assuming compression and pre-emphasis are not used) other problems can occur.

Some systems offer only SMPTE frame accuracy (40 ms at 25 fps), which is just not sufficient. Other systems allow or require the use of a computer system to provide a visual reference to find the edit points, while others offer scrub wheels akin to tape rocking.

SCMS, the anti-piracy protection code can also be a pain in such circumstances, preventing you dubbing your own tapes, let alone anyone else's.

The main factor here is that it requires an extra place to store the information, which usually means at least one extra machine, and for crossfade edits, often two, unless a digital memory is used to store the small edit section.

 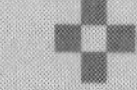

INFO

The beauty of digital tape is that changing the set is as easy as changing the tape. You can also have backup copies on spare tapes, although you will need two machines to make it, or mix it on to each tape (i.e. twice).

Digital tape does however offer fast accurate searching, digital quality (debatable) and lossless transfer (for backups, etc.). So if you can afford it, it could be the way to go. With multitrack digital systems you also have the opportunity for quadraphonic and surround sound, or extra tracks on one tape.

Digital recording

Modern systems are tending to steer away from tape (analogue or digital) because it is a linear medium. That is to get to point C you have to wind past point B and that takes time. It therefore means that the order of

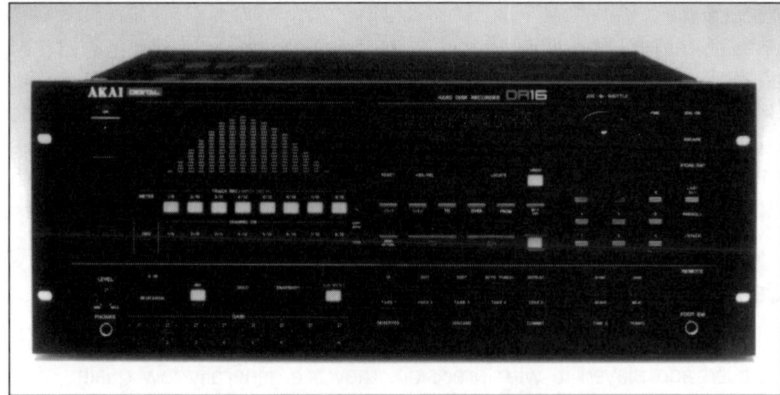

The Akai DR16 hard disk recorder offers instant random access to pre-recorded material

events on the tape is essentially fixed in terms of live playback. Time slip and crossfades between sections on tracks are also very hard, as the tape cannot physically be in two places at once.

With the decreasing costs and increasing number of computer based systems, hard disk tapeless recording is becoming standard and relatively cheap as no precision transport components are needed, except the hard drives, and these are getting cheaper all the time.

The advantage of hard disk recording is that it offers instant random access to the material, as the head just moves to the correct place. As computers have quite a lot of memory and spare hard disk space, it also means that complicated edits involving many bits of material can be performed in a non-destructive manner (retaining the original versions).

Computers of course, bring with them a new set of problems and learning curves. Modern systems offer plug and play installation and 'point and click' mouse control. Except for some very cryptic icons (little pictures which are supposed to represent a function with images rather than words) they work quite well.

Other than the fact that computers are more prone to power fluctuations and can crash (usually due to some software conflict), they needn't be any more unreliable than any of the other microprocessor devices we have come used to. So long as they work they are great. Fault finding one or installing new software or hardware can be a nightmare and there's very seldom any such thing as a five minute install. But apart from that they are here to stay.

To store CD (16 bit at a 44.1 kHz sampling rate) quality audio (without compression) the system uses 5 MB per mono track minute. That means that a 500 MB hard drive will provide 100 minutes of mono audio. For stereo this figure halves (50 minutes) and for eight track multitrack work it would reduce by eight (12.5 minutes). There is little to stop the addition of extra (or larger) hard drives, and 1 GB (1000 MB) are now affordable.

Special hard drive models often called AV drives have appeared, which are optimised for the constant data transfer required by audio applications, as opposed to computer data which is required in shorter more spasmodic bursts. These reduce the problems with the drive's own internal thermal recalibration due to heat, and can provide continuous data as required. This really helps reduce the burden on the computer system and is more likely to produce trouble free usage free from odd quirks that appear to happen for no reason.

Having random access for live use is an obvious benefit, and as a computer system, the visual cue list and locators are usually part of the system. They also often offer automation in terms of level, EQ and effects and can provide snapshot or continuous automation.

There are too many systems and versions to go in to in any detail in this book. As we've said, they are very visual and you get pretty much what you see. You just need to check that it can do all the things you want:

1. Can it do them simultaneously? For instance does it do EQ and reverb at the same time or on an exclusive basis.
2. As supplied – the machine specification may not provide all the facilities you want. As an example some features may be available only on higher specification machines (i.e. more memory or faster processor).
3. Less obvious factors. For instance it may record only on two tracks simultaneously, it may not chase sync, and varispeed and timestretch may not work.
4. There are always bugs in a system, how quickly will they sort them out and how much do they charge for upgrades?
5. Are all the features you need real-time or do they have to be done offline? Processing time is a real turn off and a disaster for live use.
6. Does it use compression or any other factors that may affect you in the long term?
7. Will it interface to the outside world happily? You may need to link it via SMPTE or MIDI (MTC or song pointer), can you remote control it, can it control other devices (i.e. lighting or relay contacts), does it have enough simultaneous inputs and outputs, what about adding your own external effects?

Tapeless production systems

Also as well as for use as a tape playback system, the tapeless hard disk recording system is also excellent as a sampler substitute, providing almost unlimited virtual memory from hard disk (which is cheap compared to the much more expensive and limited RAM memory.

With some hard disk recording systems, the tracks (via MIDI notes), and certainly always the songs (via MIDI start commands, MTC or song pointer), can be triggered remotely over MIDI, by, say, the keyboard player. This means that you can provide extra virtual sampled sounds for the keyboard player.

In terms of song production, hard disk recording offers the fine manipulation, copy and repeat, and song arrangement required in the same style as a sampler, which cannot be offered by tape alone.

For instance imagine needing to make up a composite track of short 20 second sounds from a CD. With tape, the CD would have be triggered manually the required number of times. With hard disk recording, each element can be recorded once and then copied and arranged almost instantly.

Micro time shifting, editing and quantisation are also possible, as usually are options like timestretch (making a recording fit a tempo or time slot without affecting its pitch), normalisation (optimising the level after recording) and so on. One thing that is not often offered on a tapeless system is that of real-time varispeed, which is a real shame when working with differently reference tuned instruments like piano (which may be in tune with themselves but not to A = 440 Hz), and for effect, such as to thicken vocal overdubs. Of course some digital tape machines, don't offer real-time varispeed either.

Data compression

Compression is a technique for reducing the amount of data to be handled. In digital systems this means we can reduce storage space, transmission time and in some cases do faster processing.

The problem is that no real-time compression system is lossless, which means they all affect the signal in some way. This is usually (hopefully) undetectable in normal use. However when a series of compression systems (or the same one a number of times such as internal bouncing) are used, compression can have severe side effects on the quality of the signal.

Backups

The biggest nightmare in any computer system is backups. With digital audio using quite large amounts of hard disk space, the problem is magnified. Once you have filled up a hard drive you have the option of buying another one, deleting it or backing it up. For live use we would need as many hard drives as we need for a simultaneous set. Plug-in hard drive cases/bays are available if required. There is no time for hard drive backup retrieval at a gig.

If you work with different bands or have different sets (and don't forget requests and changing the set order on the night), then that many hard drives may become impracticable. If you don't need all that data on the one night then the hard drive data can be backed up to special devices called streamers. Some systems can even use the digital input/output of a conventional audio DAT machine, although these are in theory less reliable and are normally restricted to near real time backups.

Data streaming can be at data, rather than audio, rates and provides a more secure system of data verification and parity reconstruction (a computer technique to check and restore data). Data streamers unfortunately cost a lot more than the hard drive, but once purchased the tapes are less than £10.

They do still take a long time to backup or restore. Our 500 MB drive, with a real-time audio DAT would take 1 hour and 40 minutes (100 minutes) to backup as each track has to be done individually. A data streamer operates at around 10 MB a minute, so would take about an hour (50 minutes). Of course the same time is needed to restore the backup.

Audio DAT

Note that a backup using an audio DAT is not a simple case of recording each track of audio as normal. The DAT also needs to store the cue list information which records where the file starts and stops (there may have been silence either side originally), playback details (i.e. rate) and any cue markers and edit information.

MIDI

MIDI is a very significant development in musical equipment terms. Standardised in 1983, MIDI stands for Musical Instrument Digital Interface. In simple terms it is a set of sockets on various items of equipment which lets them talk to each other.

The most common application for this is the MIDI sequencer, which can record musical performances. Extra processing can then be applied to this data before playback.

Sequencer functions include:

- Instant copying and repeats
- Transposition to different musical keys
- Individual note and parameter editing (right down to micro time shift and accents)
- Quantising – putting a performance into strict mathematical time (drum machine feel) or with later systems, humanise and groove quantise are provided for deliberate feel and humanisation.
- Automation of muting, level, panning and other MIDI controllable parameters
- System exclusive – storage of parameters that make up a keyboard sound patch
- Tempo variation without pitch change
- Near instant song arrangement and order reconstruction
- Overdubbing (even of controller information such as sustain or pitch bend) and automated punch ins
- Playback is instant random access.

TIP

Many people describe the MIDI sequencer as a wordprocessor for music. Of course if you haven't used a wordprocessor either it won't help you, but if you have it might. To my mind the spell checker is a scale correct function and word justification is like quantisation. Just a thought !

Hopefully it will be appreciated what benefits are offered by using MIDI as a playback system. These include instant random access and first generation sound. We'll discuss the application of MIDI as an automation system in the next chapter.

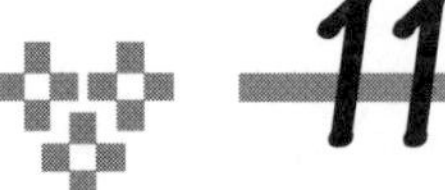

11 Automation and MIDI

INFO

The most common application for MIDI is the sequencer, which can record musical performances. Extra processing can then be applied to this data before playback, but as we'll see later, the sound engineer can apply it to automation of live sound as well.

Intro

With so many sound sources available, it is easy to run out of hands. Automation can help to reduce the burden on the engineer, leaving you free to concentrate on the art of balancing. We take a look at MIDI systems for automation and feedback eliminators.

MIDI

Standardised in 1983, MIDI stands for Musical Instrument Digital Interface. In simple terms it is a set of sockets on various items of equipment which lets them talk to each other.

Advantages of MIDI

- First generation sound – lossless recording, transmission and storage
- Instant manipulation like copying and transposition
- Event editing and manipulation is possible
- Transposition without event duration change
- Tempo change without pitch transposition
- Instant location and random access
- Near instant song order rearrangement (without using a razor blade or dubbing)
- Can control other MIDI equipped devices directly, such as mixers, effects equipment, lighting, pyrotechnics and stage properties
- Can synchronise other devices using Song pointer or MTC without using a SMPTE box.

Disadvantages of MIDI

- Set-up time and variability. A short sentence but you can spend hours trying to recreate the original, even if all the gear is available at that time.

- You need all of the original equipment to play it back to sound exactly the same.
- You can't record non-MIDI performances directly (such as vocals, guitar, saxophone, real drums, sound effects) although these can be recorded with an associated sampler or hard disk recording system.
- It's based on a computer system which doesn't recover from crashes and power fluctuations instantly (you have to reset and reload software and data). You can use a UPS (un-interruptable power supply) to reduce this risk; this could be a wise investment for a PA system.
- Quite bulky and fiddly requiring a monitor and multiple connections (i.e. SCSI, multiple MIDI connections, dongles).
- You can't extend the capabilities of equipment such as layering multiple devices, using an effect optimised for each pass several times, increasing polyphony by using multiple passes, and generally having a permanent instant record of what happened, as is possible with tape.

MIDI devices

It is important to realise how far MIDI has extended from the traditional MIDI keyboards. MIDI equipped devices now include:

- Instruments – keyboards, MIDI retrofitted guitars, wind controllers, drum and percussion pads, drum triggers (pickup mics), pianos, and even MIDI microphones exist (although in their infancy) to convert monophonic acoustic sources (i.e. vocal, saxophone) into MIDI notes.
- Virtual reality controllers – cameras converting movement to MIDI data (controller or notes), light harps (aka JM Jarre), theme park triggered sounds (i.e. rail ride contacts), MIDI gloves, and security device triggers (i.e. mats, beams, infra red sensors).
- Effects equipment: MIDI controlled graphic equalisers and filters, parametric equalisers, reverbs, delays, multi-effects, and MIDI controllable compressors and gates.
- Audio equipment – MIDI equipped mixers, VCA automation boxes.
- Other devices – lighting boards, pyrotechnics, stage properties, slide projectors, multimedia computers and animation generators.
- MIDI control pads – MIDI foot pedals, mixer tape control pads with sliders and buttons, and MIDI trackballs.

MIDI layout

The MIDI controlling equipment (sequencer or MIDI control pad) can be located near the performer or the engineer, depending on whether the keyboard player needs or wants control. The engineer really needs local control of the sequencer if it is being used for PA purposes other than

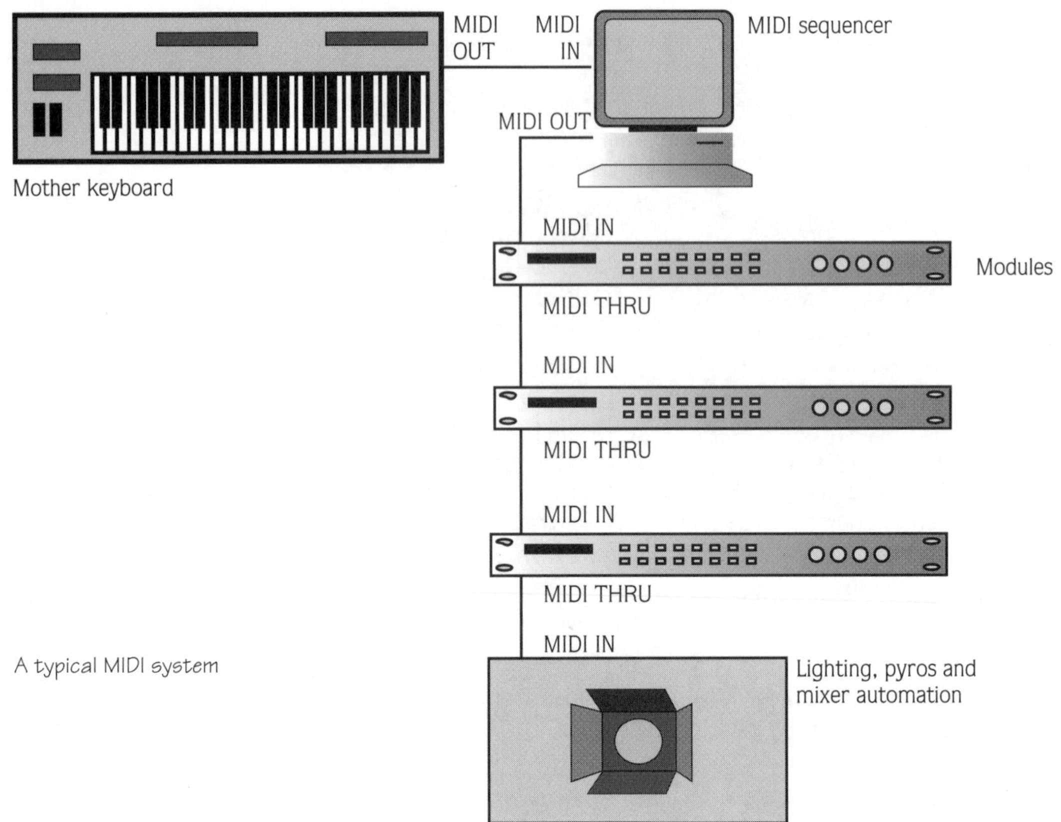

A typical MIDI system

music backings. In such cases if necessary the two sequencing systems can be linked with a MIDI cable, or in a lot of cases may be controlling completely different sets of equipment, in which case separation will be preferable anyway.

Nevertheless it is useful to have some MIDI tielines built in to the stage box. Three should suffice for any system as only an IN and OUT are required to operate all the MIDI gear. A third is suggested as a spare and for independent sets of equipment perhaps located remotely near the performers or engineer.

If you're careful about disconnecting both ends to avoid phantom power, you could even use adapter leads to convert the existing mic tieline sockets (use pins 2 and 3 only for safety). Microphone cable is ideal for MIDI – twisted balanced line with a shield.

MIDI is a 5 V digital signal and is fairly immune to cable losses up to about 40 m (120 feet). Over this distance, so called MIDI line drivers are available which can drive for much longer distances using a special device at each end of the cable.

Remote control

For live sound, the main benefit of MIDI is in remote control of equipment and it provides a cheap and fairly universal method of control and management. So let's look at the MIDI chain to see how to connect it all.

The MIDI chain

There are three types of MIDI socket:

- MIDI IN – receives data – like an ear
- MIDI OUT – transmits data to another device – like a mouth
- MIDI THRU – takes the data arriving at the MIDI IN and repeats it as an isolated feed and sends it on like a MIDI OUT – like a parrot who listens to what you say and then tells everyone else

A typical MIDI system is shown. You'll notice that an OUT is always connected to an IN, but an IN can be fed from an OUT or a THRU, depending on whether it is from the first unit in the chain or not.

The rest of the system is connected in a daisy chain of THRU to IN, THRU to IN. This can happen around five times before the possibility of a small serial delay can creep in. In such a case, a device called a MIDI THRU box can be used to split and buffer the IN to several outputs, providing a parallel connection from that point.

MIDI merging

In most cases these connections will perform every function we need. In the rare situation where we need two devices controlling the same destination, then some intelligent data management has to occur.

MIDI is a serial digital code and if interrupted produces the wrong information at the other end. This is why we can't just plug MIDI in to a parallel strip. The device to merge the two data stream intelligently is called a MIDI merge box and costs around £100.

A number of decisions have to be taken in the merge box, including what happens to MIDI clock and MTC (ignore both or accept which input), and what happens to systems exclusive which can't be interrupted very often, and the priority of controllers over notes. Sort that out with a bit of wire !

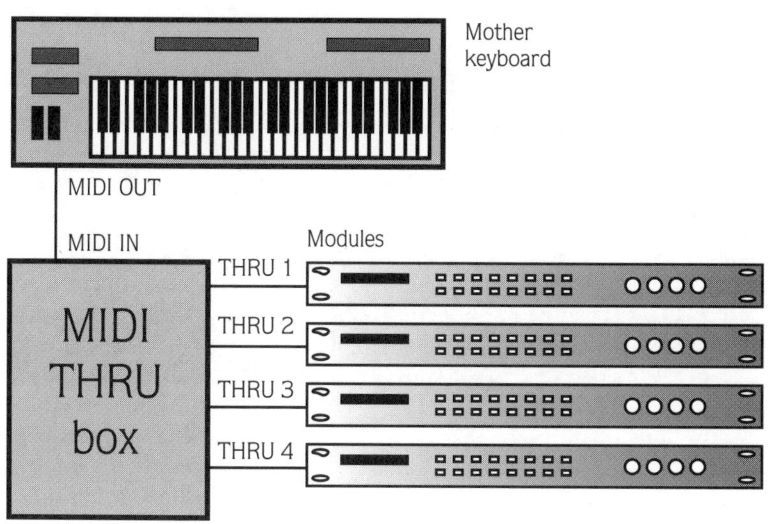

A MIDI THRU box is a wise investment for a system of any complexity. The THRU box star network prevents MIDI delays and buffers each spur

Merging really becomes an issue only when there is more than one controller, such as two keyboard players or a drum pad using the same modules or samplers. Other devices, like MIDI control pads, SMPTE sequencers and the MIDI sequencer itself, have merging facilities built in where they can combine their own data with that incoming.

MIDI codes

Some sequencers use numbers instead of names, to indicate controller messages. PC Publishing's companion volume *Music Technology Reference Book* (ISBN 1 870775 34 1) includes a full MIDI specification (as well as lots of other information). The most popular ones are listed below:

MIDI controller numbers

	Hex	*Decimal*	*Notes*
Program change	Cn pp	192+n pp	As used for scene changes in mixers/effects
Bank change MSB	Bn 00 mm	176+n 00 mm	To allow more than the original 128 patches.
Bank change LSB	Bn 20 ll	176+n 32 ll	Follow with program change in that bank.
Controller change	Bn cc vv	176+n cc vv	See controller code below for values of cc for volume, etc.
Pitch bend	En ll mm	224+n ll mm	Centre rest = En 00 40 (224+n 0 64 in decimal)

where:
n is the MIDI channel starting at 0 (0 = MIDI channel 1)
pp = program number starting at 0 (valid range 0 – 7F [0 – 127])
cc = controller number starting at 0
vv = value
ll = LSB
mm=MSB

MIDI controller numbers (cc)

	Hex	*Decimal*
Volume	07	7
Panning	0A	10
Sustain	40	64
Modulation wheel	01	1
All notes off	7B	123

Hex/decimal conversion chart

hex	decimal	hex	decimal
0	0	8	8
1	1	9	9
2	2	A	10
3	3	B	11
4	4	C	12
5	5	D	13
6	6	E	14
7	7	F	15

(hex uses base 16 rather than base 10 as in decimal)

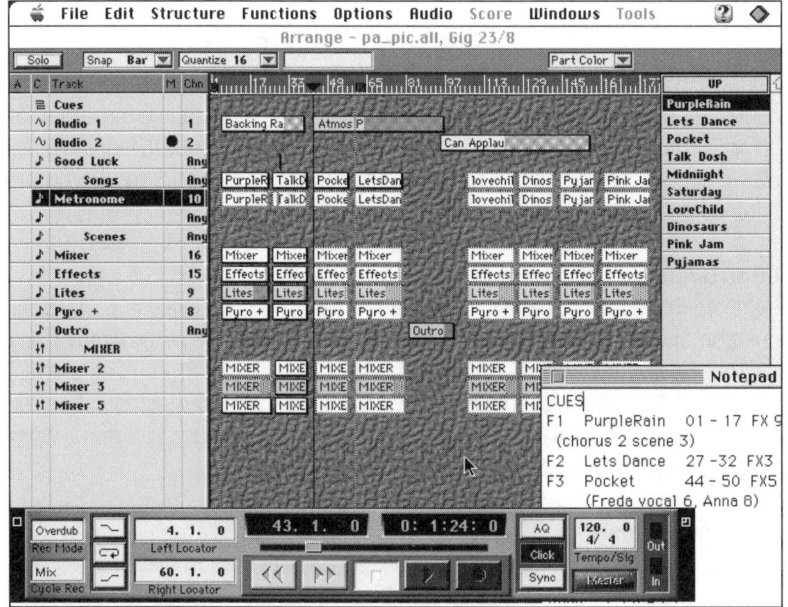

The Arrange screen of Cubase, a widely used sequencer package

MIDI concepts

The most important thing to understand about MIDI is that it is digital data about a performance. These days that performance equally applies to the faders on a mixer or lighting board as it does to the notes on a keyboard. This data can be used in real-time (as a real-time controller) or stored and played back (sequencing and mixer automation).

Although MIDI data is often used to control sound (and even trigger it in terms of a MIDI keyboard), it does not contain the sound itself. The sound comes from a device the MIDI sockets are connected to, *never* from down the MIDI cable itself.

MIDI can even be used to describe a sound in terms of each bit that a sample is made of, but not in real-time as the sampler does when reproducing the sound of a sample.

MIDI information

There are various types of MIDI information. Notes are the most obvious type. The other keyboard controllers such as the pitch bend wheel, modulation wheel and sustain pedal probably come next.

Program change

Of more interest to the PA engineer is the program change command (sometimes called patch change). Program change is used by keyboard players to change the sound of their keyboard and modules remotely. This allows the sequencer to change the sound automatically at the relevant point, or for them to call up a different sound in a remote module from their keyboard live.

In the same way, the PA engineer can call up different effects patches and mixer scenes using the same command. Again these can either be sent live, or pre-programmed as a series in a sequencer. The other significant thing about patch changes is that one patch number can be assigned to control a number of effects, each patch change then calling up a set of effects. This assumes that all the devices share the same MIDI channel, or else a sequencer (or suitable controller) should be used to send multiple patch changes on the required channels.

Controllers

Next significant is the MIDI controller. This is the equivalent to a fader position and can be assigned to control pretty much any parameter of any device. Again this can be done remotely, so you could control a MIDI graphic or MIDI controllable amp or VCA from a remote position. For the single operator/musician this can be a major boon because you don't want loads of gear cluttering the stage by your side.

Systems exclusive (SysEx) is a special type of information used to describe the parameters involved in a patch. It can be used to control a device in real time, although controllers are better suited for this if possible because they are shorter and more easily handled by readily available software.

Systems exclusive

Systems exclusive (SysEx) is a special type of information used to describe the parameters involved in a patch. It can be used to control a device in real time, although controllers are better suited for this if possible because they are shorter and more easily handled by readily available software.

The important thing for the sound engineer is that you can also store the parameter memory or mixer scenes of your device using systems exclusive. This can be saved as a separate song in the sequencer (or storage software) saving lots of money and clutter from using RAM cards. So even if someone reprograms your equipment, or it loses its memory, or you just hire one occasionally, you can reinstate it with your own preferred set-ups.

Sequencers

MIDI tends to revolve around sequencers. But a sequencer can be used to store a lot more than just musical notes. The sound engineer can use it to control intricate sets of commands and controllers for equipment, including mixers, effects, some amplifiers and other devices such as lighting and pyrotechnics.

Significant features of the sequencer for the sound engineer include:

- Visual indication of song/data position in bar/time readout or graphical song position indicator or object cue (i.e. text message)
- Multiple tracks for easy data management
- Track and pattern naming
- Instant location to bar positions using memory locators
- Muting memories can be recalled on the fly (or solo can be used)
- Graphic indication of controller positions (i.e. mixer maps)
- Saving of data set-ups to disk including systems exclusive storage of device memories
- Overdubbing allows complex operations to be built up
- Parameter editing to fine tune to perfection

The only problem with sequencing in a live situation is if the band doesn't play to a sequencer themselves. If they do the two can be locked together in perfect sync. If the band doesn't use sequencing and prefers to free run, having pre-programmed sets of events can be less useful. However in terms of snapshots, such as mixer, EQ and effect settings for each song, and short one-shot events which can be flown in manually, like fades, effect changes or triggered sound effects, it can be a real boon. With one command all of these parameters can be set instantly – offering speed and accuracy.

It is highly recommended that you learn how to control your sequencer using the QWERTY keyboard hotkeys. The last thing you want to do is fumble with the mouse which someone has probably nicked anyway!

MIDI backups

With any computer system, it is essential to have backups. Assume a worst case scenario like what happens if the internal hard drive crashes, and you can't go far wrong. Do you have the operating system on floppy and a working copy of the sequencer program, and what happens if you forget your dongle or key disk or someone steals it. And of course then there's the song data ! Well no one said it would be easy but it will hopefully be worth it !

A real pro would have a parallel backup computer running, or in most real cases a cassette of the audio and a hard copy printout (paper list) of what buttons to hit in a panic.

A lot of the technology is crossover, and if you lead a sheltered life you may turn up to the gig and the lighting man says – 'No, it's all right, our computer will do the sound thanks !' Phrases like 'the early bird catches the worm' and 'stay ahead or stay in bed' spring to mind.

Allied devices

With live performances needing to become more and more stunning, the integration of sound with lighting, Virtual reality stage triggering, computer multimedia, video walls, back projections, slide projectors, pyrotechnics and stage properties, and real-time computer animations and visual generations, it may pay to have a grasp of the bigger picture.

Ideally there will be a person in charge of each technology, but it certainly wouldn't hurt to be globally aware. There is also the problem of making it all work as a cohesive system. Communication and planning are crucial if such a system is to run smoothly.

Automatic feedback eliminators

Another form of automation comes in the automatic feedback eliminator, helping to free the engineer from one of the most time consuming processes – avoiding feedback.

By converting the sound into digital data, the automatic feedback eliminator is able to analyse the sound and process it if necessary, e.g. removing any feedback frequencies. Being digital it can do this filtering very precisely indeed (within a few hertz, regardless of the frequency) and is hence much finer than even a third octave (31 band) graphic equaliser.

Reaction to the first units has been mixed, with some people swearing by them and others at them. It depends on the type of problems you are having. The moral is try to borrow or hire one before you buy. If it does work for you then it will really help you to get more level and avoid feedback, and you'll be able to concentrate on other things such as mixing.

12

Problems and solutions

Intro

Due to the transient nature of a live performance, even if the venue hasn't changed from one gig to another, many of the conditions will have. In this chapter we take a practical look at avoiding mains, RF and lighting interference, avoiding feedback and coping with changing acoustics.

Interference

In any system, external interference from other devices is always a problem. For live sound, as well as often having very transitory systems with a new venue and almost a new set of problems every night, we also have one of the worst possible scenarios – miles of cable and reproduction at high volumes. There are three main ways of picking up interference:

- Radiation – air born
- Induction – distance induced
- Cable born – i.e. mains feed

Radiation

Air born interference is almost impossible to avoid. Heavy metal screening (or an earthed cage) would stop it, but the only practical way is to treat the source by fitting suppressors to the units themselves. These are special capacitor and transient type electronic components which dissipate the energy before it has a chance to radiate. They can be fitted to light switches and motors and so on. Maplin Electronics and Radio Spares (Electroplan) do these sort of devices, though as mains power is involved I recommend that they are fitted by a qualified electrician.

Of course once you're at a gig there's little you can do about it other than stopping it happening (i.e. don't use that light or fridge please Mister) or trying some alternative orientations and distances. Other than that we have to try and live with it.

If this interference is being picked up by the microphone cables, it can be reduced by using the slightly more expensive star quad type cables, although I've not seen this implemented in a whole multicore stage box cable. Keeping cable lengths to a minimum may help, and try to avoid coiling them and turning them in to loop aerials.

Using screened cables is mandatory in these situations too. Some types of cable are better than others in this respect. Tin foil types offer better protection than single braid, because they are less likely to open up with use. Double braid screens can offer similar results, but vary from manufacturer to manufacturer. Conductive screened cables are not as effective at all frequencies as their foil counterpart but are sometimes used on their own or to support a braid type cable. Although these cables are cheaper, lighter and more flexible than foil types, this has to be weighed against a one off cost and performance. No doubt some cable manufacturers will disagree, and in some individual cases they may be correct, but you have to be aware of the potential problems. It is very uncommon for a supplier to say they'll take it back if you have problems with it, so unfortunately it is *your* choice.

Radio mics

Air born interference also includes radio mic problems, such as dead spots and spurious channel pick up. Diversity systems placed as close to the desired source as possible will help. You can only test different frequency units at each venue and hope. As we've said, it can pay to have a cable backup ready to go. Ensuring that your system uses legal frequencies will stop you getting a heavy fine and help to avoid problems as that's partly why the set ranges were introduced. New systems shouldn't be so near the CB radio ham and taxi cab frequencies since the tighter regulations were introduced.

Induction

When two items are close together, they can induce interference in each other through capacitance and inductance effects. Cables are usually the culprit here, although some rack equipment just doesn't like living together in the same rack. In such a case only distance will help. Don't confuse this with ground loop hums which we'll discuss under ground loops in a minute.

To prevent inductance between cables you should use only high quality screened cables (although the screening is more for radiated than induced interference), using a balanced system where possible. Cables should not be coiled, and wherever they cross a higher power cable (i.e. mains, lighting and computer cables), should only cross at a 90 degree right angle. This minimises the area between the two and hence reduces interference. Distance is of course preferably, but there is only so much floor space. If you can run down a different side or section of the venue to the lighting, etc., then that would be favourite.

Cable born

Unless you've paralleled your cables together with the lighting ones, the usual source of cable born interference stems from the mains power cables.

The same kinds of noise sources as for radiated air born interference can also transmit down the ring mains and via your power cable. This type of interference is much easier to deal with. Using a separate power ring will normally alleviate most of the problem (so again don't plug in to the kitchen), but for, extra protection, mains suppressors can be used. These are little boxes containing mains filtering and transient suppressors and attenuate this type of interference.

In extreme cases you may even want to consider using power conditioners and regulators. Computer system users will usually gain similar benefits from using a UPS (uninterruptable power supply – as strongly advised for live use with computers) anyway. These devices will also help to cope with mains power fluctuations (10% dips are quite common) which can play havoc with a system.

TIP

Distance is best but if you have to cross a power cable do it at 90 degrees.

Mains and phases

A good clean separate power feed is essential. This usually means just using a separate ring main socket that no one else is using. Where a three-phase power system is available, this can provide added protection. It is beyond the scope of this book to discuss using three-phase mains as they need to be handled by a qualified electrician.

You should always opt for a separate phase to the lighting and kitchen/air conditioning. Be very careful *not* to mix your equipment between different phases at any point. This will hopefully be impracticable, but make sure it cannot happen. A three-phase supply carries much more voltage on it than a normal house supply, and you or your equipment may fry. If you are responsible for cooking a member of the public or band through such negligence you will not have a happy life even if you do have public liability insurance (which I sincerely hope you do).

Ground loops

When connecting any quantity of equipment (even two devices) the chances of ground loops occur. Despite what anyone may tell you *NEVER DISCONNECT THE MAINS EARTH* from a piece of mains equipment. It is potentially dangerous and can be lethal. There should be a better way to solve such problems.

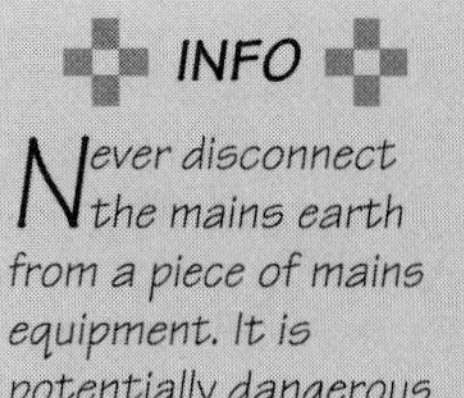
INFO

Never disconnect the mains earth from a piece of mains equipment. It is potentially dangerous and can be lethal.

People say that it is OK to remove mains earth so long as one piece of gear is earthed, and usually they opt for the mixer. This is *not* the case. It would mean that although the equipment may be at earth potential, the earth system is relying on a few strands of audio cable screen to carry 13 A of current if (and when) a fault develops. Anyway, as soon as the audio leads are disconnected, the case is potentially live.

Earths are provided for protection, not just for fun. They prevent any metal parts of the case from becoming live should a fault develop and so prevent electrocution. Do not solve your audio installation problems by removing the mains earth. I may not even be able to tell you 'I told you so' because you might be *dead*!

Conventional ring main

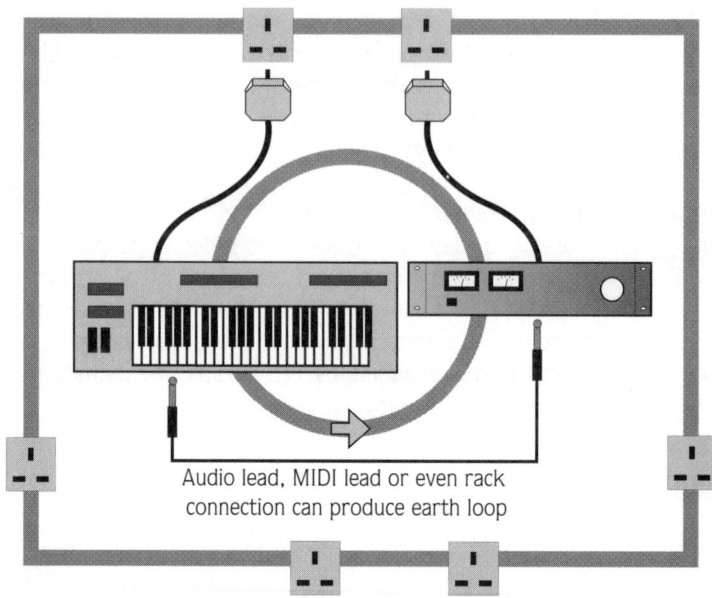

So how do you avoid and solve ground loop problems then ? Firstly all connected equipment must come off the same power point, using a system of short lead mains blocks, all fed from one other. This is another reason for siting all the gear together. This provides the most common earth point possible. Be very wary of cheap mains blocks. The connections they provide can be quite high resistance and not provide very good earths for audio purposes. They may buzz out fine with a resistance checker, but you need less than 1 ohm in total to ground for an audio system. This common plugging will remove 60% of ground loop problems.

First you need to determine if any noise is due to a ground loop, equipment noise, a missing earth, or to some other cause such as induction. You really need to start with a bare system and connect one bit at a time starting with the amp and speakers.

If you have a hum, does it go when you unplug the power cable instantly or does it linger while the power drains away ? If it goes instantly then you know it's an earth loop. In this case you need to look at the audio, and data if any, connections. I've even had MIDI leads which caused an earth loop on a keyboard audio output, and a hard drive SCSI cable causing problems on the audio outputs of a tapeless system. You need to see if removing the input cables removes the problem. If so you can treat those. If only removing the output cable solves it, it might be just because you can't hear it (so double check it's not induced).

To treat an audio cable, it's easiest to make an adapter lead which has only the live connections connected and the earth lifted at one end. On a jack or phono plug this means lifting the screen cable which connects to the body housing of the plug – usually the biggest tag. On an XLR this will mean pin 1 (they are numbered on the plug). For a DIN plug this is the very middle pin (and sometimes a case clamp) which is confusingly called pin 2 (it may be numbered).

Mains star distribution system

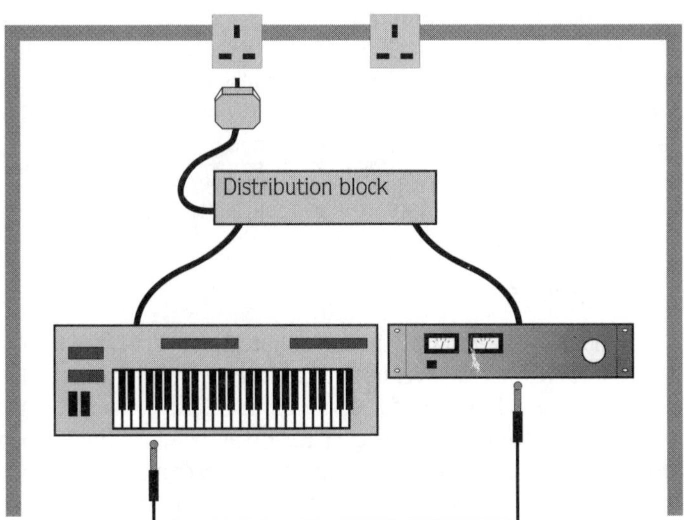

With a star system no earth loop is possible

Remember you may need to do this to both channels, and you may need to try other connectors as well, such as inputs and any auxiliary connections. If this cures the problem then you have two options. To use a lead with this arrangement, or to modify the equipment socket, or mark it with gaffa tape to remember to use this earth lift adapter. Each has its own merits, but for mobile PA use you want the fastest and most flexible system (which is probably the adapter).

The alternative to the lazy approach is to use a resistor in series with the screen which has a higher resistance than the ground loop. This normally equates to between 100 and 600 ohms. This then offers the added protection of supplying a conducting screen which will screen better than a shield connected at one end only. The choice is yours. Again for mobile PA installations you may need to increase the resistance value as the ground loop resistance of a new venue may be higher.

If the hum doesn't go as soon as you unplug the lead, it may be induced or just being picked up on the inputs or other connections. You will need to remove the input leads to see if it goes. If it doesn't then you will need to investigate any adjacent equipment for induction – power amps often induce in to effects processors in the same rack for instance.

Another favourite is ground loops caused by the metal housing of the equipment sharing a rack earth. In this case isolating washers and spacers are the answer.

If disconnecting the earth and removing the input and taking the unit out of the rack and holding it in thin air still doesn't work, then maybe the unit just hums on its own. You could also try adding extra earths as outlined below as perhaps its own ground path is not good enough (i.e. dodgy mains block).

Ground loop or ground off ?

If none of this seems to work, it may be because of a completely different problem – a missing ground. This is usually most easily found by having an earthed bit of wire and touching it on supposed earth points, such as chassis and socket screens. If it cures it then you have to find out why the earth is missing (a dodgy lead which may get swapped in position later ?) or add an extra earth if there is no logical reason.

Loudness is a relative thing, and if you can't move the threshold up, then you need to move the noise and ambient threshold down – if you can.

Sound globes

We should also mention at this point interference from the sound level limiters installed in many venues by the local council. Although these often seem to be set a little low for some types of performances, they cannot simply be bypassed by plugging in to a wall point in the kitchen. These globes set what is considered a safe and environmentally friendly sound level, effectively constituting a bye-law active in that venue. By bypassing these devices you are effectively breaking the law and potentially endangering the hearing of the audience (and the good will of the neighbours).

Also by using mains sockets so far away, you also introduce extra risks of poor quality mains, with spikes and power dips on it caused by heavy machinery and appliances such as fridges, ovens and dish washers. The chances of mains hum caused by ground loops also rears its ugly head.

Lighting

One of the biggest sources of audio interference tends to come from lighting circuits. Dimmer racks often radiate RF SCR interference, and, by switching large lamp current, induction is likely.

Most lighting problems are caused by adjacent cables, so use distance or cross at right angles. Star Quad microphone cables may help, as may better screened (tin foil) cables and multicores.

Some lighting problems will be radiated, in which case you've got little hope other than distance and angle and maybe a notch filter. Other lighting problems may be mains born, either noise or power dips. Double check there isn't a cleaner ring main feed or, as mentioned before, conditioners (effective but expensive) and filters (cheap) are available.

Feedback

Feedback is almost a whole book in itself. It is probably the number one problem with live sound. We've all had it. We could get exactly the sound and level we needed if only the thing didn't howl.

Feedback occurs when a sound from a speaker is picked up by a microphone (or similar, i.e. turntable or acoustic guitar pickup), re-amplified and then this process is repeated until that howl appears. With strong sources, such as loud singers, the problems are fewer, but there are many occasions when you need to add gain, which will make the system become unstable and just waiting for feedback to occur.

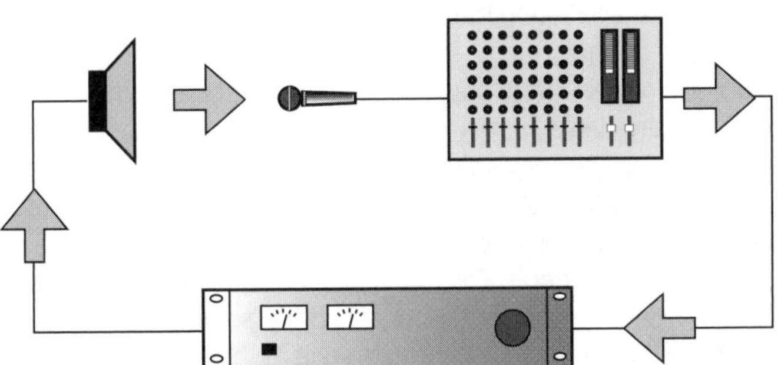

The feedback path – on each cycle the sound is amplified until it ends in a high pitched howl. Ringing occurs just before, so it should be avoidable. The cures are, reduce overall gain, reduce gain at feedback frequency i.e. with a graphic EQ, and position equipment so rear of microphone faces speaker front.

Surprisingly feedback does not occur at the particular frequency of the sound which triggered it. Feedback occurs when the system is excited and it then occurs at its own resonant frequency. This will vary depending on the venue and the equipment used. It occurs at the speed of sound 330 m/s (1100 feet/s) which is why it is a continuous howl, rather than repeated sounds. Then once started, the feedback is a continuous cyclic stream, which isn't dependent on the distance between the microphone and speakers to determine the feedback frequency (pitch). This is actually governed by the dominant (there could be several) resonances of the system.

Any system which has positive coupling between its input and output, and that means any PA, can become unstable when the amount of amplification (gain) or the degree of coupling exceeds the critical point. In our case the coupling is the placement between the microphones and speakers, which is why pointing a microphone at a speaker makes it howl. The amount of amplification depends on how loud the sound needs to be at the venue and how much gain has been introduced in the system to achieve it. This is where our expression gain before feedback comes from. Any help the speaker efficiency (and dispersion to an extent) and microphone sensitivity (not to mention the performers themselves) can give us, the less unstable our system will tend to be.

Microphone choice and technique

The dominant resonance of the system will depend on the natural resonance of the microphone and speaker diaphragms. Unfortunately moving coil microphones (our most common type) are often resonant around the 2 kHz mark, an important point for intelligibility, and one we don't really want to cut with a graphic. Ironically the least popular live mic, the ribbon mic, offers a natural resonance which is towards the higher end of the frequency spectrum, thus reducing the possibility of feedback quite considerably. This is why some PA engineers swear by the Beyer M260 and M88 ribbon microphones for instance, which have also been designed for more road ruggedness.

Capacitor microphones often have a natural resonance at around 8 kHz which is quite useful both sonically (for a clean sound) and for feedback purposes. If necessary a slight treble roll-off can control any feedback. Because of their cost and need for phantom power they have not been

It is a lot easier (and more natural sounding) to remove a frequency than to try and boost one that may not even be there. That would only result in a coloured sound, noise and a greater chance of feedback.

very popular in PA applications. But are becoming more common as prices fall, they become more roadworthy, and more PA mixers come with phantom power fitted as standard. Separate external phantom power units are available in any case, using battery (with voltage multiplying techniques) or mains versions.

Microphones which colour the frequency response themselves can be useful. For instance a presence peak means the electronic gain at this frequency can be reduced (or not boosted). Similar ideas occur for the bass boost caused by the proximity effect of some directional microphones. However if this boost isn't required it is better to avoid it rather than have to equalise it out.

Close miking techniques and the use of hypercardioid and supercardioid microphones also help to reduce feedback. With such directional microphones you have to be careful that the sound from the rear lobe doesn't colour the sound too much, but overall the tighter pickup pattern will offer more advantages than not.

Positioning

We can of course control the positioning between the microphone and speakers quite easily, so long as the performers can restrain themselves from roaming or standing in front of the speakers, at least for direct sound.

It is worth mentioning here that even sealed speakers still have an output from the rear of the cabinet, so even with the rear of the microphone facing the rear of the speaker (as desired) there is still the chance of feedback. The speakers should therefore be placed forward and to one side of the microphone and ideally the speaker front should be angled slightly away from the microphone.

For reflected sound (reverberant field) the situation is not as clearly controlled. The shortest path for reflected sound is actually from the rear of the stage. This can be reduced with the use of an appropriate acoustic absorber. In our case the feedback frequencies are likely to be in the mid range (around 2 kHz) so drapes and curtains (ideally air spaced to the wall by at least 15 cm (6 inches) will help here, and won't look too out of place. In theatres these are quite often used as standard.

The low frequency response can also be cleaned up by using 'suspended' wood panelling on any bare brick or concrete surfaces. These act as small bass absorbers, reducing the reflected sound at these frequencies and cleaning up the reflected sound quite considerably.

The curtains should be a heavy type using thick folds and run from floor to ceiling and across the whole of the rear platform from left to right. For extra attenuation, the back wall could also be treated with acoustic tiles or rockwool to attenuate the remaining sound after it has gone through the curtains (before being reflected back by the wall). This treatment can be confined to the more central area which has the least curtain in line with the sound (reflections from the side walls will direct the sound at an angle through the curtain folds).

A theatre backdrop cloth is not highly absorbent and quite reflective of high frequencies once thick painted with the scene. Hopefully the HF reflections will not be troublesome. The backdrop should be brought forward from the wall by 30 – 60 cm (1 – 2 feet) and then extra absorption placed behind it to dampen the wall reflections further. Large pillars and the ceiling above the stage itself may also need to be treated.

For low frequency feedback (growl rather than howl) then a design for this frequency range will need to be considered. Other factors may also contribute to these effects, including standing waves from the room design, so they may not just be feedback related.

Acoustic guitar pickups seem quite prone to triggering these kind of problems, which are reinforced by the guitar acting as a soundboard for the feedback too. Similar effects can occur with structural transmission through flooring and mic stands, which in such a case should be isolated with rubber mats.

Speaker positioning

We have already discussed placing the speakers away from the direct line of sight of the microphone to avoid feedback. What is not so apparent is feedback caused by the reflected sound which can also be controlled by speaker positioning and tilting.

By angling the speakers away from the side walls and tilting them down to the rear of the audience, rather than the rear wall, we can help to reduce the reflected sound. The audience make fairly efficient sound absorbers, and providing the seats are occupied or padded they will not reflect the sound either. Any raised bare seating platforms will obviously introduce unwanted reflections.

The speaker tilting of column speakers can be judged quite well by eye. In the front seat your eyeline should be in line with the bottom of the speaker, and at the rear seat to the top line of speaker. For raised or balcony seating, similar guidelines apply, but the eyeline and tilting is reversed.

If two sets of speakers are used because of the depth of the venue, then the pair closer to the rear wall should be angled inwards more to the opposite corner to stop strong reflections.

Summary

- Speakers should be angled and tilted towards the audience whilst avoiding strong reflections from large surfaces such as rear and side walls.
- Drapes or acoustic treatment can be used to avoid reflections from the rear of the stage.
- Microphones should always face away from speakers.
- Speakers should be placed in front (of the rear of the microphone) and to one side and angled slightly away from the microphone, for maximum feedback rejection.

Other techniques for feedback reduction

Graphic tuning

Another technique for controlling feedback is to equalise it out electronically. This is most commonly done with the graphic equaliser. You should

Feedback frequency plot shows how a resonant frequency can be treated with a graphic equaliser. Feedback often occurs at more than one frequency so you may have to repeat for each problem band

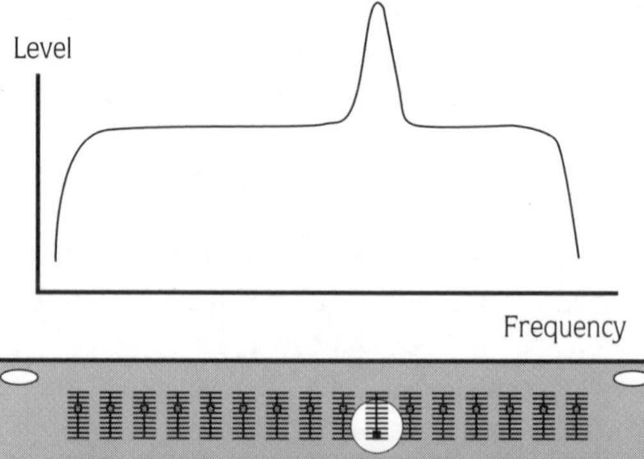

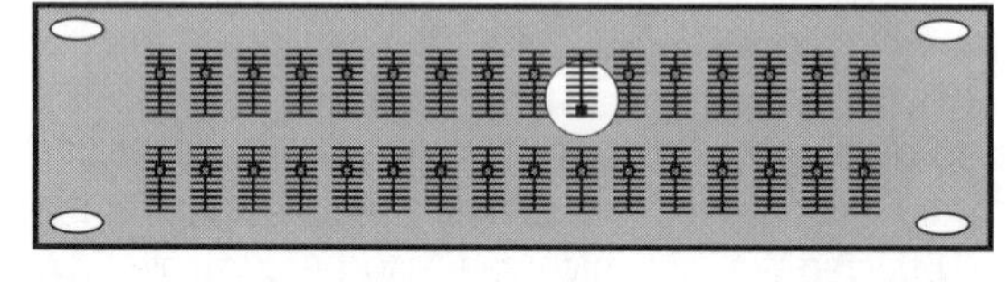

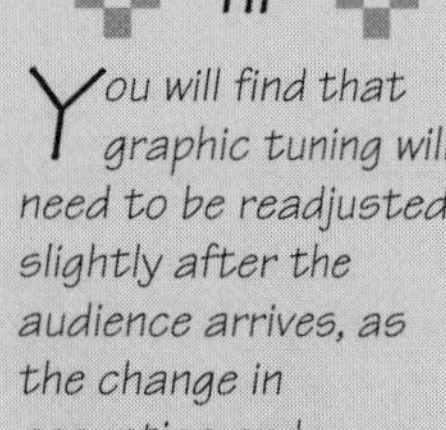

TIP

You will find that graphic tuning will need to be readjusted slightly after the audience arrives, as the change in acoustics and temperature will both affect the sound.

use a 31 band type. You may also have to use two sliders to tame the required frequency. To use the graphic, increase the level to just where feedback starts to occur (ringing). Then turn down each graphic slider to see if it stops the feedback. If it has no effect, return the slider to normal. If it does, then set it to reduce by a small amount (say 4 dB) until the feedback stops. Then increase the level. Feedback may occur at the same frequency or it may occur at the next peak, in which case that frequency needs to be found and the procedure repeated.

With experience it will become possible to choose the frequency by ear rather than empirically.

As a guide, here are some common frequency ranges for different types of feedback.

Feedback/frequency table

Feedback sound	*Frequency range*
Hoot	250 – 500 Hz
Singing tone	1 kHz
Whistle	2 kHz

Feedback is unlikely to occur in the 8 – 16 kHz or 16 – 62 Hz ranges, so you should try these faders last.

Notch filters

An alternative to the graphic equaliser is a device called a notch filter. This uses a very fine band stop filter which can be tuned in to remove a very narrow frequency range, often in the order of a few hertz. Some units have two or more of these in series so that the next dominant frequency can also be treated. As the frequency range affected is so narrow, the effect on the music is almost negligible. They are also comparatively cheap.

Another advantage of the notch filter is that as well as attenuating the feedback frequencies it will also help to attenuate the reverberation reproduced through the system which often occurs at around the same frequency. This can help to improve intelligibility quite noticeably.

The notch filter works by superimposing an exact opposite of the feedback resonance peak. All systems have peaks and dips in their response, and speakers can easily deviate over a 10 dB range. In amplifier terms, 10 dB is a lot of power. If the speaker has a 10 dB peak in its response which is causing feedback, then we have to reduce the amplification by 10 dB or get feedback. By using the notch filter to remove the peak we can get our level back again.

The graphic equaliser can be used to do this, but as we have discussed, even a 31 band graphic has quite a wide range associated with each slider, so it may affect the music considerably. None of the sliders may even appear at the problem frequency, and it is quite common to have to use two adjacent sliders to control the required 'in-between' frequency. This of course affects the sound of the music even more.

The notch needs to be at exactly the same frequency and phase. If it is phase displaced by half, then there will be no attenuation, only a slighter sharper side. At a quarter displacement, the peak will only be reduced by about half. If the notch is too narrow (a higher frequency), then two smaller peaks are introduced instead. Increasing the depth of the notch just produces a deeper well between the two peaks.

In short, the notch needs to be the same frequency and phase, depth (equal and opposite level) and width as the peak we are trying to alleviate, or its effectiveness will be vastly reduced. With the graphic equaliser the chances of this are slim.

Another factor is that the response peak will not be just at one frequency and will normally be at a rate of 6 dB per octave around the centre frequency. Some graphics allow this response to be used via a switch, whilst others are fixed at 12 or 18 dB per octave response. Any notch therefore needs to match the range or width of this peak as well, which is where the concept of notch width comes from.

Pitch shift

Other units which have been used to reduce feedback include pitch shifters and delays. The pitch shifter applies a 5 Hz shift in frequency, on the theory that, as what comes out isn't exactly what went in, feedback shouldn't occur. In practice it only helps to a degree.

There are also other factors to consider here, namely the effect on the performers. Because our perception of pitch is logarithmic rather than linear, each octave has a different amount of frequencies in it.

For instance the low C octave goes from 262 Hz to 524 Hz while two octaves up it is between 1048 and 2096 Hz. So 5 Hz is a more significant change at the lower octaves than 5 Hz in the higher octaves.

For the octave below this (131 – 262 Hz), C is 131 Hz while C# (C sharp), a semitone higher, is 139 Hz – a difference of 14 Hz, making 5 Hz a third of a semitone (33 cents). For the octave below that, 5 Hz becomes greater than a semitone (C = 66 Hz, C# = 69 Hz – a 3 Hz difference).

For speech (300 – 3400 Hz range) pitch shifters work well as the percentage change is small and does not affect the material. For music, as the pitch shift is constant, it will affect low frequencies more – as we have seen, by up to a semitone.

So if it is to be used, it needs to be applied to higher range and vocal signals only, leaving out bass guitar and low synth sounds. Inserting one on the insert points of a sub group dedicated to these instruments is one option.

The advantage of the pitch shifter is that it needs no setting up (assuming the factory calibration remains stable) and isn't affected by the venue acoustics.

Pitch shifters have a circuit which includes a 5 Hz oscillator which is used to multiply different phased versions of the original which are then combined in special ways. The result is an output of the original signal shifted by 5 Hz. But, as most things to do with sound are not linear at all, frequencies and the phase change circuitry inside may not be perfect at all frequencies, and this may introduce beat frequencies which although small, will colour the sound.

Delays

Delay lines offer very small feedback improvement. Although in theory they should delay the original and prevent feedback, in practice, once initiated, the feedback occurs at a steady rate. Also the natural reverberation of the venue acts as a type of reservoir, and so the sound is still present in the delay gaps. Even with long delays, which would be impracticable to perform to, the feedback still occurs.

A modulating delay can offer some improvement, but it also introduces pitch changes. There is of course a limit to the settings in terms of performing with them and the improvements are still quite small.

Gating

It is worth mentioning here that a gate will *not* help to reduce feedback. This is because the gate is only closed shut during silence, which would reduce feedback. However as soon as the gate is triggered open by any sound, it becomes effectively transparent, thus providing no feedback aid at all.

13

Live sound in allied industries

Intro

Sound is all around us, the need for quality audio is not unique to the gigging musician. In this chapter we look at theatre sound and the conference and exhibition industry.

Sound of course appears in all sorts of places – not just rock venues. So sound systems are found in all sorts of places: conference halls, restaurants, in-store entertainment, theme parks, museums and themed museums, exhibitions, multimedia presentation systems, games arenas, in-flight entertainment, and of course the dreaded lift musak – which was once described by one shopping addict as playing 'party music which makes you feel like you're having a good time'.

In many of these applications the role of the sound system is to reinforce the original and present the sound to a wide number (or in some cases a selective number) of people, rather than to present something larger than life and impressive through sheer volume.

The theme park could be our exception here, as it may be desirable to create a sonic impact to supplement a visual one, and this may rely heavily on sub bass feelings and panning HF. There is usually of course the need to overcome quite a lot of ambient noise from machinery and people, which has to be weighed against the potential close proximity of the audience/visitor.

Ironically in many cases, the sound sources will not be from microphones, but from electronic sources such as tape, CDs, computer soundcards and CD-ROMs, and even musical instruments such as samplers. Also often the source and the speaker system can be a long way away as there is no need for their visual association.

This means that there is less chance for feedback of course, which is one of the bugbears of the live music system. The irony is that in this situation we may not need this type of level capability.

The sound system design can vary dramatically for each application and may consist of numerous separate feeds (including multi-lingual from multitrack tape machines) to multiple low level speakers well spaced and with a deliberately limited coverage area. Some of these installations may even be mobile, such as touring buses or floats often with wireless systems feeding them from a central sound studio.

In these cases, remote (or local) control of volume is highly desirable, and with the cable lengths involved, may well use the 100V line system for distribution, or alternatively local amplifiers with balanced line feeds to them.

In these installations, interference is even more of a problem, as many lighting, computer and electro-mechanical systems may be involved, each with its own breed of problems. The sound system may even need to interface to a number of these, which may require isolating/matching transformers (a DI box for line level to line level use), and ground loops are almost a certainty with the multitude of power systems and the lengths of cables between them. You should be aware of three-phase supplies in such installations and seek the help of a qualified electrician if in any doubt. The use of a mains isolating transformer may be advisable if no firm answers (or an electrician) can be found.

Power conditioning in such venues is also a necessity as the mains power is very unlikely to be that stable.

All the guidelines already discussed in this book apply to all of these venues. The only difference is the mechanisms used to deliver them, as the requirements are so different for each venue. Hopefully your 'common sound sense' will get you through. A wireless talkback system is probably essential so that you can personally experience the other end of the sound system, rather than rely on the hope that the system is properly set up and isn't distorting, etc.

Conference systems

We'll take a quick look at systems for business here, which is rapidly expanding from the conference room to multimedia presentation systems designed for the public.

Rifle microphones are another possibility for difficult venues which are ambiently noisy. However a number may need to be involved to get a sufficiently large hot spot to prevent the speaker swaying off mic. An ambient mic may still be required to pick up the audience questions and feedback as well.

Microphones

It is common for microphones to be as unobtrusive as possible and hence not hand held or stand mounted. Lavalier microphones are one option as they give a close stable pickup even with movement, and can be clipped to the speaker's lapel and forgotten about. Because of the cables involved, they are not ideal where there are multiple speakers, or speakers who need to be mobile to use a white board or slide projector.

A wireless system can be a solution here, but it is an expensive option if several mics are required. Hot swapping the sound system is as unacceptable in the middle of a presentation as it is in the middle of a performance.

The PZM boundary microphone is another popular choice, as it can be attached to a surface such as a lectern or table top and ignored. It will have a wide field of pickup and is suitable for groups of speakers. It has a hemispherical pickup pattern so some care (or baffles) needs to be used to remove extraneous noises and the rustling of scripts. Perspex mounts can be a lot more visually acceptable in applications which need them.

PZMs are even found in police stations, presumably because the whole wall becomes part of the microphone, the confession can still be recorded, even from the floor – *just joking ozzifer.*

Surface mount microphones are an unobtrusive solution for conferences

Loudspeakers
In this application what we are after is an even spread of sound rather than an overpowering central focus. Multiple small speakers can be used (using our 3:1 spacing rule – speaker to person or greater). In fact the intimacy of such a system can be a positive advantage in this application. You may need to provide headphone feeds for multi-lingual applications.

Amplifiers
For business purposes we can fairly safely apply our 1W per person guideline and maybe even halve it. Depending on the size of the conference we may need to employ 100 V line systems, but usually with this size of venue a house system will already be installed which can be patched in to from your own system.

Multimedia

Multimedia is just a buzz word for the combination of sound, graphics and computing in one environment – as in multiple media. Multimedia is being used for all sorts of applications including:

- Entertainment (games systems, 3D television, and interactive television)
- POS – point of sale as in sales kiosks and car rental terminals.
- POI – point of information as in public information in town halls, maps and museum catalogues.
- CBT – computer based training as in self education/instruction programs and distance learning terminals.
- Living catalogues – such as catalogues for car parts, fashion model databases, and clothes and hair modelling packages that let you see yourself in that situation.
- Virtual reality and modelling – where you can experience a virtual world or product before it is manufactured. Although sound doesn't play a major part in these experiences currently – it should.
- Presentation systems – as used in theme parks and interactive multimedia experiences.

You can achieve a vast improvement in sound quality from a computer system by using conventional external speakers powered by a dedicated amplifier, either built in as a powered speaker or as a stand alone system.

The speakers 'bundled' with most computer systems are usually very low grade and meant for individual use rather than by a group or couple. Most soundcards have a provision for either dedicated line outputs or at least a jumper to reduce the gain (and noise) of the on-board soundcard amplifier. Some modern soundcards even have provision for a digital output which can then be fed to a higher quality DAC (digital to analogue converter) or for use in a digital sound system.

Shielded speakers

Another thing to be wary of in computer environments is that the monitor screens are susceptible to the magnets in speakers which can distort the picture. Magnets are also not good news for computer media, such as floppy disks and hard drives, so care should be exercised in placing, to avoid such mishaps. Some manufacturers can supply so called AV versions, so named because of their use in video environments which have similar requirements to computers. They have magnetically shielded cabinets which are more suited in these situations. They are not usually that much more expensive.

Security and safety

With any situation involving people and especially the general public, it is of prime importance, both professionally and legally (and financially through compensation) to make sure that the sound system is secure and safe.

All equipment and cables must be secure and not cause a hazard to anyone. You cannot rely on people seeing a cable or head height obstacle and bypassing it – equipmnent should be out of reach or marked clearly. As the public will be in such close proximity and unsupervised, double precautions must be taken in terms of anchorage and prevention against theft or tampering.

14

PA in action

Intro

One of the biggest problems with live performances must be the venue. Here we take a brief look at some of the physical problems of venues.

Planning

Because you never really knows what problems you will encounter at the venue before you get there, good planning and a working system is your best ammunition. It is surprising how much you can forget in the heat of the moment otherwise.

It is highly advisable to do a site survey the day before to plan access and work out any special requirements. Gather as much of the following information as possible to ensure the gig goes smoothly:

- The number of performers and instruments
- Details of the backline amplification
- The stage and performer layout
- The foldback monitor requirements
- Audience and stage location and sizes
- The size and position of the audience and the level required
- Any special feeds required for recording or back stage use
- The location of power points and any other site equipment

With this information you can properly plan for microphone choice and mounting, cable layouts, number of speakers and amplifiers, auxiliary equipment and backup provisions. This will reduce the number of conflicts and go towards a smoother gig.

As well as the site survey you will need to do the following:

- Prepare any inserts (tapes, MIDI, etc.)
- Test the gear and your backup provisions
- Create and follow your checklists. It's amazing what you can forget in a rush

Ideally the mixer position should be about midway in the audience position so you can really hear what the audience is hearing.

The setup

When you arrive at the venue for the gig, you should set up as follows:

1 Locate and interconnect the equipment

The amplifiers and auxiliary gear should be situated next to the mixer so that you can monitor and adjust them easily. The gig could involve cross patching faulty equipment and you don't want to have to fight your way through the audience or musicians to get to it.

When rigging it is essential to identify the microphones so that you can distinguish them from the monitor position. Performers have a tendency to move around without telling you. The use of coloured windshields, cables or coloured tape is recommended.

You should also try to group your cables and multicore runs and mixer channels between instrument type and location. Having stage left mics on the left of the mixer as well really helps. The mixer channels need to be grouped by instrument as well, so all your drum mics should appear on adjacent channels.

2 Test all the equipment and run it up with a tape you know

This will also help to warm up the amplifiers and reduce the strain of a full load from cold. You'll also get a good idea of how the venue is going to sound and how well the speakers are covering the area, whilst you set up the rest of your equipment.

3 Adjust the equipment in a logical order

Equalise the main outputs

You should start by equalising the main outputs. You can do this in a number of ways. One is by equalising your known tape and the other is to use one of the microphones in front of the speakers and increase the gain until feedback starts, then find the problem frequency and reduce it. Don't introduce wild amounts of cut or you may affect the tone of the music as well. Usually adjustments of less than 6 dB are called for. Then repeat the process until you're happy.

Adjust the coarse gains and set approximate levels

You can set the coarse gains from ambience or the renowned voice check. You will soon get to gauge what the likely level sensitivities are going to be between instruments.

Set vocal microphone levels

As the vocals will be the loudest thing through the PA and will cause the most problems with feedback, you should start with these first.

- Adjust the instrument levels
- Set up the stage monitors and equalise them for feedback

Run through

Run through each instrument separately and then with the full band. Set your balance and get sufficient level for the venue. You will probably need to re-equalise your graphics and set the channel equalisation for each source. This will be a compromise between getting the sound you want and avoiding feedback. You will tend to have more success if you try using equalisation cutting, rather than boosting all the time.

Adjust amplifiers

Adjust the amplifiers for the level required for the audience. Note that the monitors and microphones should not be moved once set. Planning and rehearsals are obviously essential for a smooth performance, after all, this is live sound and there are no retakes.

Multiple sets

It is wise to make a worksheet for each song in the set, to remind you of any special equalisation or level changes, or effects settings or cues. If you're working with multiple bands the worksheet is essential as you'll have little time to remember the settings for each band as they come on.

Troubleshooting

If you do encounter problems during the performance, you'll really appreciate having the solo button and meters so you can locate the fault without disturbing the audience further. Sealed headphones are essential to locate the problem aurally as well.

You are less likely to encounter problems if you learn the feedback levels during the rehearsal. Once you reach that point you know you've got nothing left to play with before it squeals. Your option then is to take some other action, such as reducing the volume of the other instruments or to adjust the equalisation to get more effective level through increasing the tonal effectiveness.

Once the audience arrives, you'll find the feedback threshold can go up slightly as the audience absorb some of the level and reduce the ambience and reflections which contribute to feedback.

Case studies

Classical concert

For classical music, the sound of the environment is an important part of the sound itself so close miking techniques aren't appropriate. More ambient miking techniques are used to reinforce the sound. In the case described here the microphones were used to feed a broadcast TV system.

The mic technique involved using several sets of ambient mics to cover the area. The exact positioning of the performers could change with each set and was prone to last minute changes, as camera angles and other logistics intervened. It was important to have a mic technique to cover these eventualities with the already slung microphones. In extreme cases coverage could be supplemented by extra spot mics if quick fix techniques were necessary.

The microphone technique used depended on the piece, in terms of its period, instrumentation, number of soloists and any unwanted audience participation. For instance the sound stage for a piece by Bruckner or Mahler would be different from a modern chamber work or a work from Mozart. All the subtleties of extra performers or techniques need to be transmitted by a suitable microphone technique.

Sometimes this can be judged from attending rehearsals at the hall, although the performers will be adjusting to the venue at this time, so the sound is far from guaranteed. Usually however, because of the pressures of time, this often has to be done drawing from experience or knowledge of the composer's work, or sometimes just from the score.

Sometimes the orchestra layout would be changed, with the woodwind section moving to the normal hall risers, meaning that the whole orchestra would move back 3 m from the pre-slung microphones. At other times the score required different sections to be brought in to focus and different microphone emphasis required.

The number of people in the audience will have a marked effect on the reverberance of the venue, and, as it is so integral to the sound, will dictate the balance of microphones at different distances.

Classical concert sound system

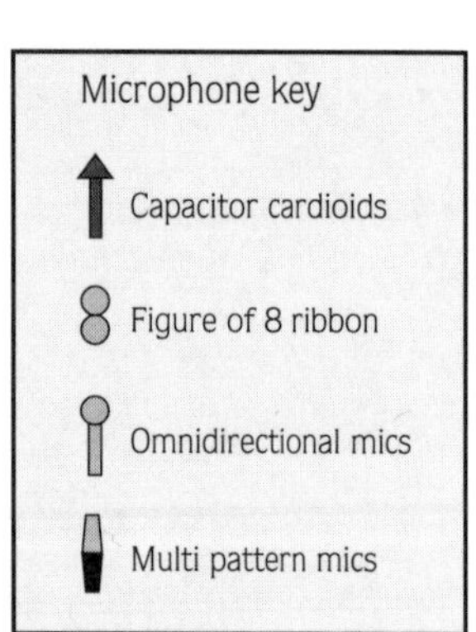

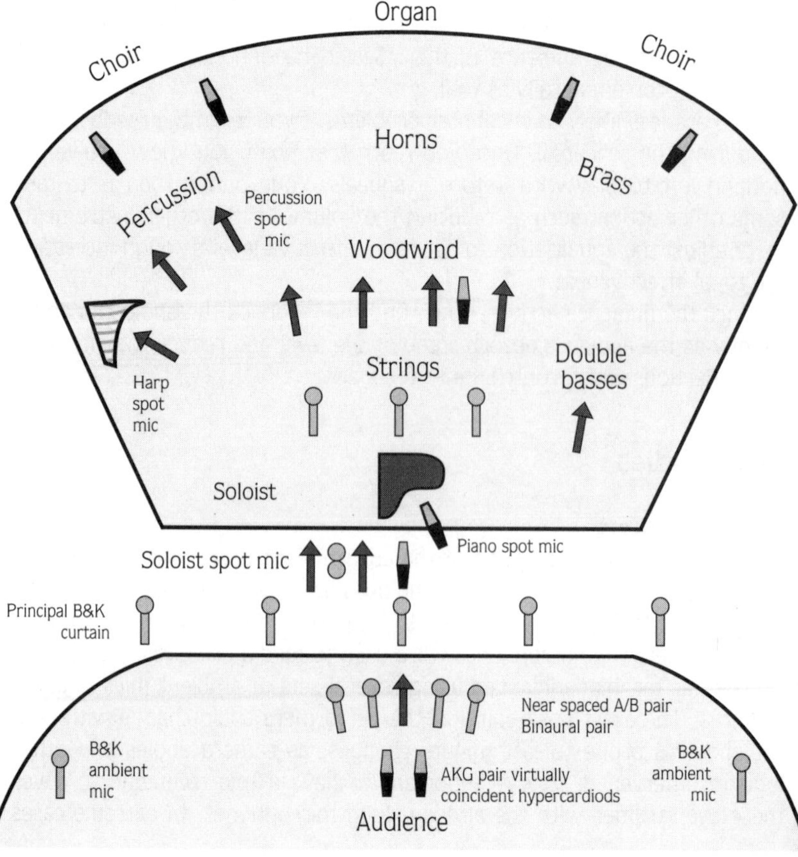

One technique is to fade up each stereo pair in turn to hear how they sound generally. Some can be rejected at this time. For instance with small groups of players, a distant sound will often not give enough detail and may need to be supplemented with closer microphones, such as the curtain of B & Ks, or for very small groups, the conductor's pair, as shown in the diagram. Judging this from an empty hall can give a completely different perception to the real situation with a full audience. Fine adjustments need to be made quickly and inaudibly in the first few seconds of the performance. Generally a sharp change will be less detectable than a change animating itself noticeably over time.

It is also wise to find different sets of microphones that will give the required sound, both for choice and in case of problems. It is also important to take a break from balancing to gain a fresh perspective on the sound, for details which you may have overlooked or grown accustomed to during rehearsals and balancing.

To reduce such a large acoustic event to a small stereo field, requires very good stereo definition, which stems from detailed microphone placement and accurate balancing.

Computer conductors

A complex percussion piece involved different tempo maps for each player, which would then combine in sync at different times. An Atari computer was used to play six separate metronome feeds at different tempi for each musician, via in-ear monitors, acting like a multiple arm conductor. Apparently the sequence took two weeks to program, so it will be appreciated how hard it would have been to conduct manually.

A number of problems arose during the performances. First, one of the stereo mics developed an 8 dB level drop probably due to a dirty connection in the miles of cables, so an alternative pair of mics was used for safety. One of the percussionists missed a cue because one of the click tracks failed to work, because the artist forgot to plug in the headphone cable; and a spot radio microphone on a mobile performer, which had worked perfectly in rehearsal, decided to pick up a lot of interference noise during the performance and had to be used at a lower volume than desired. Still that's live sound for you.

Rock concert

A famous large name band had to play the Sheffield Arena – apparently an excellent venue. The show used a T stage layout with lots of props and video walls to an audience of 12,000 people.

It was decided not to use delay speaker systems in order to retain the focus of the centre stage. 72 speakers were used as stacks stage left and right, supported by six bass bins and two six way clusters.

The six way clusters using piston drivers, were placed stage left and right and were focused towards centre stage and down towards the audience. The controllable directionality of the cluster (because of its very narrow dispersion) allowed control of spillage from underneath the PA, which can be a problem.

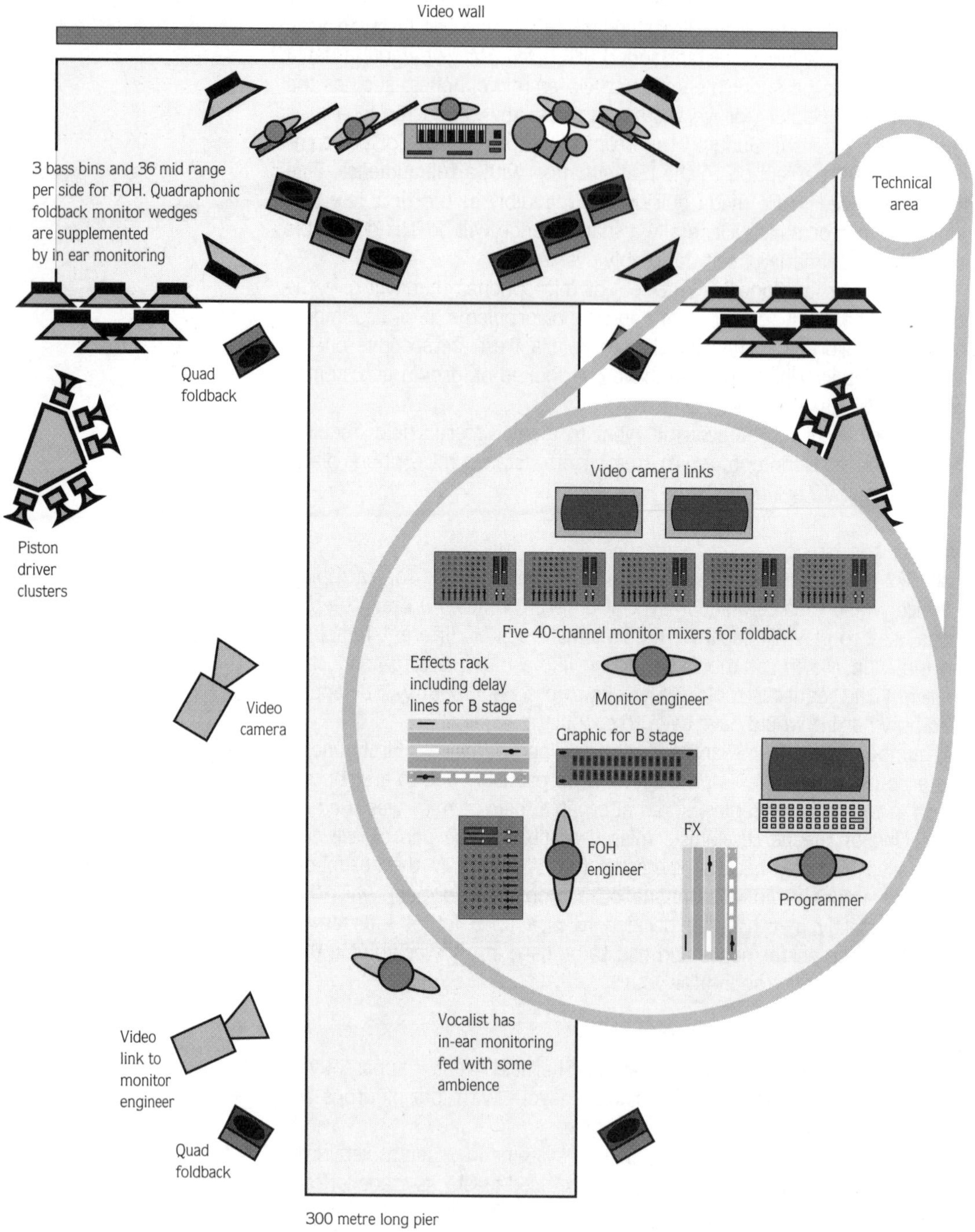

Video wall

3 bass bins and 36 mid range per side for FOH. Quadraphonic foldback monitor wedges are supplemented by in ear monitoring

Technical area

Quad foldback

Piston driver clusters

Video camera links

Five 40-channel monitor mixers for foldback

Effects rack including delay lines for B stage

Monitor engineer

Graphic for B stage

FX

FOH engineer

Programmer

Video camera

Vocalist has in-ear monitoring fed with some ambience

Video link to monitor engineer

Quad foldback

300 metre long pier

Rock concert sound system

The main desk had 40 channels each with an integral gate and compressor, six auxiliary buses, six mono sub groups which can also be used as inputs if desired, and eight stereo auxiliary returns which also featured three way parametric EQ. The desk featured VCA control, so any fader could be routed or turned in to a sub group.

The biggest problem was to try to produce the band's album sound with the same incredible detail, live. This included even having a microphone for a door bell sound effect.

To avoid feedback on the small B (centre of the T) stage, a separate graphic feed was 'ringed' out (see the section on tuning a graphic in Chapter 12) but it was still touch and go, as it depended so much upon audience noise level.

A quadraphonic monitor mix was required, including panning to the centre T stage. This was joystick controlled so that it was easy to move the entire monitor mix for each musician as they moved around the stage.

In-ear monitoring was used in addition to traditional wedges and side fills, but had to be time-align compensated as the band went down the 300 m stage, due to the delays from the main PA on the main stage. The in-ear system worked well because it provided a quality feed without the need for damaging high levels. A small amount of ambience was also fed in to the monitor mix to help the feel of the in-ear system. The monitor mix was controlled by five monitor consoles, providing about 200 channels. Audio and video links smoothed out communication between the monitor engineer and the band.

Equipment wise, offstage were extra racks of MIDI gear including five keyboards and racks of modules controlled by a dedicated programmer naturally. A rack mounted computer acted as a live conductor and supplied keyboard sounds and visual cues. A shaker sample was used as the click track for the drummer. The computer also helped to synchronise the visual effects.

Needless to say the power was controlled by UPS's offering 40 minutes of backup power and to keep the computer equipment happy.

INFO

Around 36 headphone and upstage/downstage mixes were happening simultaneously, complete with a programmable equaliser rack. Over 40 support staff were involved in the technical support for the band.

Microphone usage table

Vocal	Samson wireless with Shure Beta 58
Kick drum	Beyer M88
Toms	Sennheiser MD421
Overheads	B & K 4011
Snare top	Beyer M88 and Shure SM57
Snare bottom	Shure SM56
Piccolo snare and hats	B & K 4007 (Omni)
Guitar	Shure SM57
Audience mic	Sennheiser MD421
Keyboards and bass	DI boxes

Cabaret

A pop trio of piano, bass and drums had to play a small stage area in an L shaped venue as a cabaret type gig. The biggest problem for them was that, as the stage was so small, it meant the piano and drums were very close together, and hence the singer originally had to shout to hear himself. By adding a local foldback system for the piano, this problem was overcome.

The main PA feed was via stacks on either side of the stage, as the audience interested in the music accumulated in this area. The rest of the venue, including the bar on the other part of the L, was supplied with a lower level feed via a link to the house background PA system. This was used for taped background music when the live performers were not on. The background PA system consisted of a number of small 4 inch type speakers, well spaced about 3 m apart across the ceiling.

The audience was around 300 capacity and was covered with some 15 inch units with horns driven by an 800 W amplifier. The inclusion of the foldback, a 12 inch with tweeter wedge driven at 100 W, provided the foldback.

Cabaret show sound system.

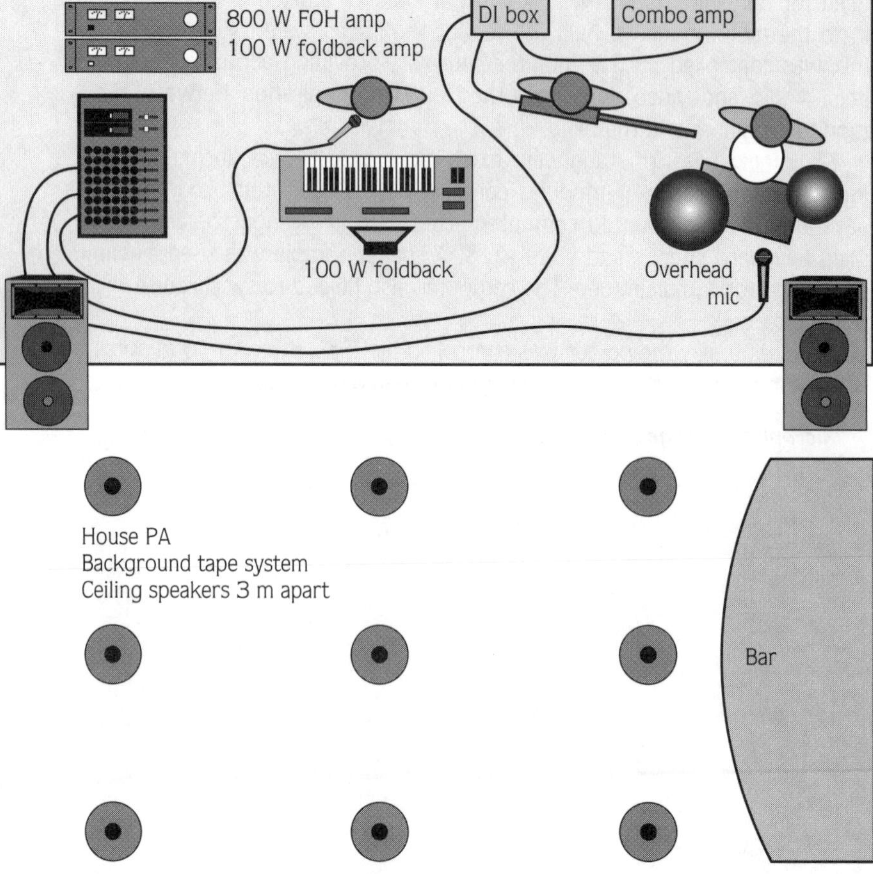

Theatre production

A West End musical production consisted of 30 actors, 50 children, a string quartet and a house band of drums and percussion, French horn, oboe, cor anglais, sax and flute, two guitarists, electric bass and three keyboard stations. A finale also involved a pre-recorded quadraphonic surround sound disco. See picture overleaf.

A couple of cabled microphones were used for the vocals, but the rest was covered by 24 radio mics and some spot mics, including three slung float mics over the pit. Because of channel restrictions, the radio mics had to be hot swapped a couple of times between the cast. Sweat from the actors was a problem with some of the radio mics, and several had to be swapped for a different brand which were less susceptible. Also at some points the mics had to be used from within some large head masks which seriously compromised the sound quality and required heavy equalisation. The radio receivers were situated by the stage, but a computer was used to monitor the radio mic status and signals (RF, AF and over modulation).

The main PA was handled by dual arrays situated in the proscenium arch. One array used dual concentric drivers – one pair per side for the stalls, and another pair higher up for the galleries. These were supplemented with some bass bins. The second array used 15 inch speakers with two bass bins under stage in line with the first array. An 'advance bar' was also used with four speakers feeding the upper circle and a speaker and bass bin focused on the front stalls. Two cannons were also flown from this bar. The mix of band and vocals could be split between the systems to provide control over spread and coverage and to match the sound to the individual characters and scenes. Meanwhile the monitor mix was via six wedges carrying the band for the actors.

The finale was covered by a separate system of 68 4 inch speakers across the seats of the venue supplemented by bass bins in the corners for the desired upbeat disco ending. A rack of six delay lines was used to align all these speaker sources together. This was fed from a digital multitrack system with the pre-recorded quadraphonic backing.

Two mixers were used for FOH, providing up to 72 channels. Some MIDI controlled mixers were also used, especially to recall some 30 scene changes of microphone set-ups as well as controlling effects programs and delay configurations. This was all driven from a computer sequencer which was manually cued. This worked by playing one bar a sequence. At the end of the bar was a *stop* command which was looped back in to the sequencer (the echo/thru function was turned off) which automatically stopped the sequence before the next set of instructions in the next bar. Text cues were also written in as part of the sequence for visual reassurance.

Some dynamic EQ units were also used, which are level threshold triggered to come in to action. This helped to reduce feedback as the performers moved near the bass bins (by cutting the LF until they sang) and for removing harshness at high volumes.

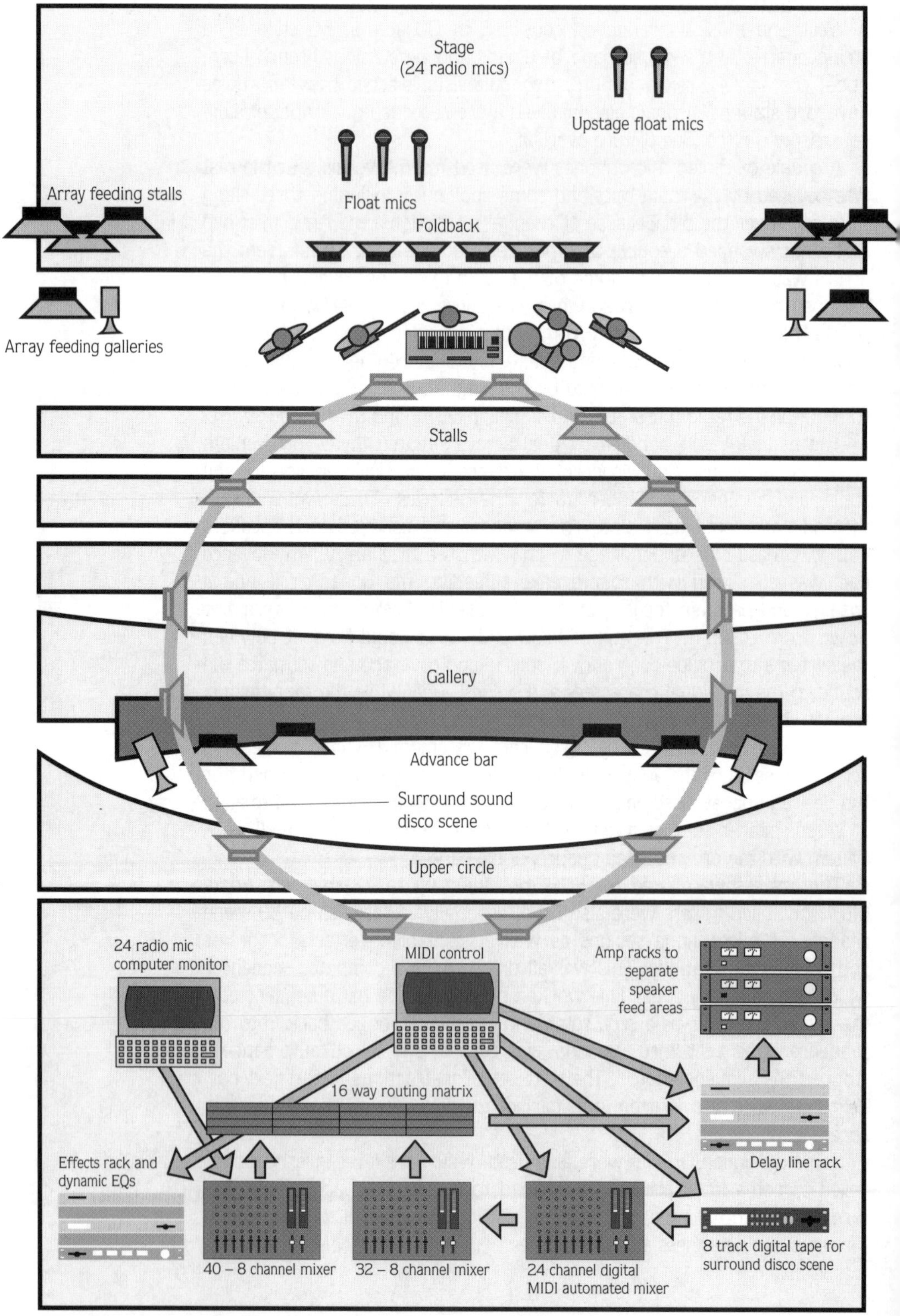

Stage
(24 radio mics)
Upstage float mics
Array feeding stalls
Float mics
Foldback
Array feeding galleries
Stalls
Gallery
Advance bar
Surround sound
disco scene
Upper circle
24 radio mic
computer monitor
MIDI control
Amp racks to
separate
speaker
feed areas
16 way routing matrix
Delay line rack
Effects rack and
dynamic EQs
40 – 8 channel mixer
32 – 8 channel mixer
24 channel digital
MIDI automated mixer
8 track digital tape for
surround disco scene

15

Safety

Intro

In any situation where the public are involved, special consideration has to be given to safety. We look at electrical and mechanical safety and hearing damage.

Safety issues are critical in live sound work because of its transitory nature and because the public are involved. Public liability insurance is highly recommended, but is still no excuse for negligence and lack of common sense, which has its own legal as well as financial implications.

Rigging and cabling

The largest problems are in rigging and cabling dangers where equipment may fall on, or collide with, or trip, a person. All cables must be secured and marked, and no cables at body or head height are allowable unless secured to a wall. A modification of the old saying could be 'give them enough cable and they'll sue you'.

Equipment needs to be highly stable with no chance of falling on to someone's toe. Anything which is slung really should be installed by special contractors who are trained and insured for such activities. It's not just a case of knowing the chain strength, reinforced fixing points and being able to climb a ladder, there are also potential structural implications of load bearing and building damage. And remember we're talking nuts and bolts here, screws can work lose.

Electrocution

The danger of electrocution is obviously high, especially when interconnecting with other people's equipment. We've already mentioned the dangers of connecting equipment powered from different phases of a three phase supply (*just don't*) but there are also potential dangers between your microphones (which may just be earthed) and the musician's own equipment, like guitars, keyboards and amplifiers.

Any reports of shock must be investigated seriously and not just attributed to static electricity and nylon socks. Obviously this needs to be investigated with caution using a neon screwdriver or preferably a volt

TIP

The worst type of shock is between both hands as the current travels directly across the heart.

meter referenced to ground. Remember that even a few volts can kill. When handling potentially dangerous equipment it is important to try to have rubber soled shoes and to use one hand only.

The use of isolating transformers can help to remove the risk of electrocution, and in some places (such as the BBC) they insist that user equipment is powered only via such devices. Obviously leaving the earths intact on all equipment is the main protection against electrocution, which is why we suggested alternative methods using the audio cable screens in our section on dealing with ground loops (Chapter 12).

Hearing damage

Another potential risk is in hearing damage. Some councils insist on the installation and use of sound globes at public venues. These will automatically trip the power if the sound level goes above a certain point. Some of them do seem to be set rather low, but are normally around 96 dB, which is regarded as a safe level for around eight hours of exposure. These should obviously not be tampered with or bypassed, by plugging in to the kitchen, for instance.

If you break these rules then you are liable for the consequences. More cases of lawsuits for hearing damage are occurring daily, with many workers claiming against their employers for not providing adequate warning and protection. There is nothing to stop similar actions against venues and their service providers (you). How easily these actions can be supported and claimed is debatable, but common sense should prevail. What the test case result would be of the drunk who insists on going to sleep with his head in your bass bin is yet to be decided of course !

House feeds

Caution also needs to be exercised in allowing for the provision of venue fire alarms and the like, and that your sound system or you have a provision to let them be heard. This can be a particular problem when tapping into the house systems of the venue. It is another element that you need to consider or take specific advice on.

Crew safety

Other than being able to deal with personal situations of violence and abuse, usually through some process of diplomacy, assistance, avoidance and self defence, you also have to look after your crew and yourself.

Weight lifting is a major problem as you tend to get a bad back for life. It is also not good to drop cabinets and amplifier racks on your foot, or trap your hands in a doorway. These are all obvious commonsense statements and yet these 'accidents' still happen.

When loading/unloading it is highly advisable to have some leather rigging gloves to help you. They will give you a better grip and give you some protection from scuffs and blisters.

You should not attempt to lift anything you don't think you can handle. The law says you don't have to. It's not a case of being a wimp, but of getting the job done quickly and safely and protecting the equipment. When going upstairs with cabinets or racks, their should be one person at the top and two at the bottom. Remember to keep your hands and fingers under the object so that you can pass through doorways. You should also carry such objects from the bottom and *not* from the handles. The handles are only supplied to manoeuvre the equipment and not to lift it over objects. It is also easier to maintain lifting something if it is part of your body's vertical column, rather than a side annexe to it applying fulcrum leverage.

Most people know the proper way to lift things, but of course it is hard to apply and still manage to lift something. You should keep your back straight and bend at the knees, which should give you the power to lift it. The other trick with lifting is inertia and breathing. You can't coax a box in to lifting itself!

> **TIP**
>
> *You are strongly advised to label all flight cases as to their stage position so that any helper can position them in the right area right away, without having to re-move everything yourself later. It also helps in developing a system to pack it all away in the truck and in distinguishing your gear from the band's.*

Attitude

A strange one to include in safety, but a confident professional attitude which mirrors who is in charge and what is acceptable will help you alleviate many problems. Don't get distracted by spurious requests until you are rigged and sound checked. The audience won't understand that the concert can't start because you were held up fixing a lead for somebody. Remember that the sound check should be under your control – it is not the time for a band rehearsal or jam session. It is your one chance to get the sound right before the live performance.

16

Sound facts

Intro

We take a look at the principles of sound in the hope that it may develop your practical understanding of what is happening.

In most instances, we regard sound as the vibration of air within the audible frequency range of 20 Hz to 20 kHz. The sound source causes oscillation in the air which in turn causes oscillation in our ear drum. Sound can of course also be transmitted through other media, including water and solids.

Human hearing, when at its peak during childhood, can detect frequency variations on the range of 20 Hz to 20,000 Hz. The abbreviation Hz is short for hertz, after the name of the man who defined the quantity of a cycle per second. The abbreviation kHz (kilohertz) means one thousand hertz (k = kilo = 1000 times).

As we grow older, our high end hearing response tends to become less sensitive, and 15 kHz is quite a common limit in 30 year olds. It is debated whether frequencies above this range have any effect or not at this age, but most tests prove that their absence still makes a difference to the listener regardless of age.

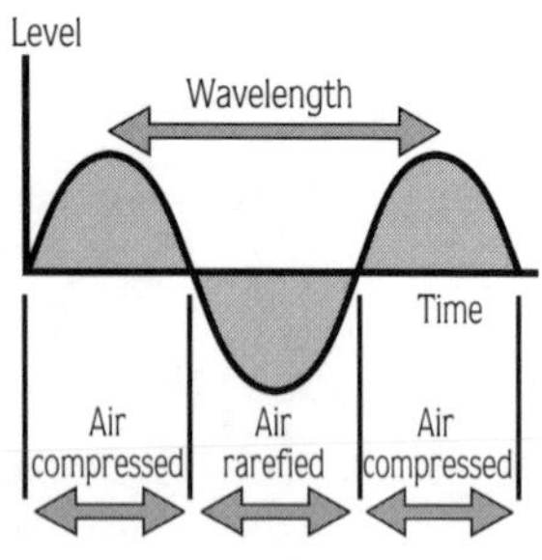

A typical sound wave showing our perception of sound. The shape of the wave indicates the harmonic content, and the number of cycles per second is a measure of the frequency

For a number of analogies you can regard sound as being like the waves in the sea, with peaks and troughs of water moving in a certain direction. In the case of a speaker this is more like dropping a pebble in a pond with the waves spreading outwards and getting shallower as they do so. In reality sound actually moves more like a piston, compressing and rarefying the air pressure in sympathy with the sound source. In either case it will be appreciated that the air is not actually moving itself, but transferring its energy to the adjacent air.

The common way of visualising sound is with the sine wave. This represents level or air pressure against time as shown in the diagram.

The diagram represents the air being at rest, then compressing and then back to the rest point, and then rarefying (moving in the other direction) until it comes to rest again. This is called a cycle. In the sine wave case the waveshape is a symmetrical and pure one, however music consists of the resultant shape of a large number of these waves, all at different frequencies and levels, with the result that it has a very complex shape.

You may remember your physics teacher using a slinky (a spring that could climb stairs) to demonstrate the difference between transverse and longitudinal wave travel (piston and waves).

The number of cycles per second is measured in hertz, and the listener perceives this as pitch or frequency. The more times this cycle happens per second, the higher the frequency, and the higher pitched is the sound we hear. Conversely the pitch sounds lower if fewer of these cycles occur per second.

The length between the start and end of the cycle can be measured as a distance and is termed its wavelength. This is related to the speed of sound in air by the formula:

C = W F

where

C is the speed of sound (341 m/s at 15.5 degrees C)
W is the wavelength (in the same units as C i.e. feet or metres)
F is the frequency measured in hertz

Using simple formula transposition we can see that

W = C / F and F = C / W

One of the most important aspects of understanding sound is the relationship between the elements in this formula. It means that as the frequency increases the wavelength decreases, and conversely as the frequency decreases, the wavelength increases. If you consider our visualisation of the sine wave, you will see that this has to be true. As a practical example, the following table shows some examples.

Wavelength/frequency table

Musical note	Frequency	Wavelength feet	inches	cm
A1	55 Hz	20	240	600
A2	110 Hz	10	120	300
A3	220 Hz	5	60	150
A4	440 Hz	2.5	30	75
A5	880 Hz	1.25	15	37.5
A6	1760 Hz	0.625	7.5	18.75
A7	3520 Hz	0.312	3.75	9.375
A8	7040 Hz	0.156	1.875	4.688
A9	14.08 kHz	0.78	0.938	2.344

(assuming speed of sound is 330 m/s)

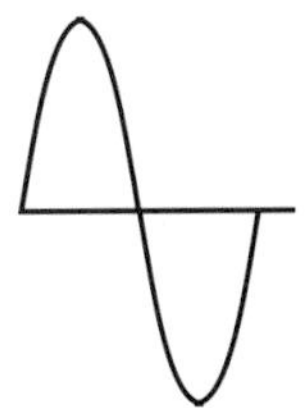

Square wave

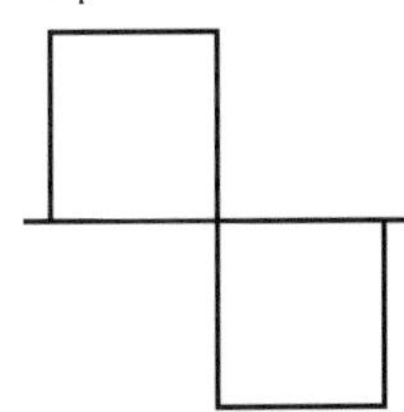

Clipped sine wave

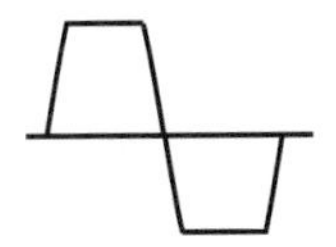

From the top:
Sine wave
Square wave
A clipped (distorted) sine wave – looks very like a square wave and can blow speakers.

In practical terms, the wavelength is of significance for reproducing or monitoring bass frequencies. To reproduce a bass frequency effectively, the diameter of the diaphragm needs to be in the order of a quarter of the wavelength. Similarly the bass response of a venue can be dictated by how much distance is allowed for the sound to develop, with a greater bass response being noticed as you go further away from the source. This may also be partly due to reflections from the wall or other surfaces, but it has an effect nonetheless.

It is worth mentioning here that the bass response is increased by 3 dB for each surface (assuming it is of significant size), which is why a speaker in a corner will be bass heavier than one positioned out of the corner. This can be used in live sound to reinforce (or reduce) the bass level if desired. It may also be appreciated why moving a microphone a small amount (even a couple of centimetres) can cause a drastic difference in the response of the high frequencies, which have wavelengths of under 5 cm (2 in) at frequencies above 7 kHz.

What we regard as the level of a sound is actually our perception of the pressure of the medium on our eardrum. This SPL (sound pressure level) is measured in decibels (dBs). As we have already mentioned, a dB is just a ratio. The reference we use in the SPL ratio is 1 pascal (or 1 newton/m^2, 10 μBar, or 10 dyne/cm^2). SPL equates to voltage in terms of its dB calculation (20 log).

SPL is the square root of intensity. Intensity of course is the amount of acoustic power, akin to the watt in electrical terms. It is expressed as a ratio to the faintest sound which can be heard – the threshold of hearing, which is 10^{-12} watts at 1 kHz. Sound intensity is not often used in sound measurement.

SPL dB table

dB	ratio
0	1
2	1.259
3	1.413
6	1.995
10	3.162
14	5.01
20	10.0
40	100
60	1000
80	10,000 (10^4)
100	10^5
120	10^6 (a million)

Decibels

We have already discussed the dB ratio measurement in Chapter 3. In sound terms though it is worth mentioning the following facts:

- The range from the quietest sound to the loudest one we can hear is a million to one and the sound intensity is the square of that (a million million).
- 120 dB is considered the threshold before permanent hearing damage takes place.
- Twice a particular SPL level is 6 dB, while four times is 12 dB, eight times is 18 dB and so on.
- The smallest difference that can be detected is 1 dB, although 3 dB is a more practical amount. 10 dB is about three times as loud,

Phase

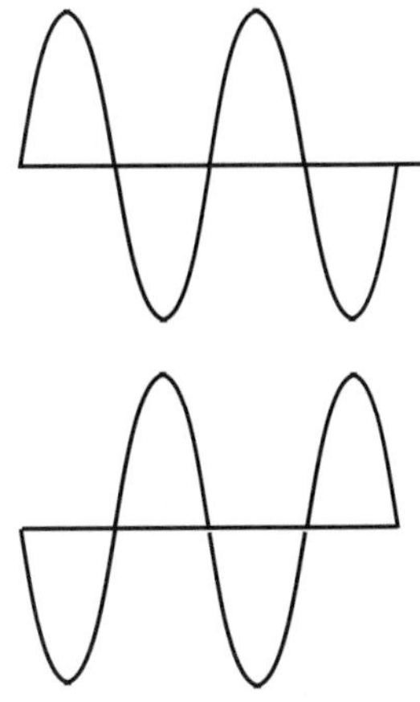

There is a phase difference of 180 degrees between these two sine waves

Another very important aspect of sound is phase. A waveform goes through a cycle of 360 degrees before returning to its starting point. Phase is the relative position of one waveform cycle relative to another. If two sources (i.e. two speaker cones) are both pushing simultaneously, then they are said to be in phase. If one is pushing while the other one is pulling (i.e. one in its positive half of the cycle while the other is in its negative half) then the sounds are said to be 180 degrees out of phase. If two signals which are 180 degrees out of phase are added then they will cancel. With sound of course, we rarely deal with pure sine waves (except test tones), so we are much likely to get phase cancellations at some frequencies and not at others. For instance the bass response is usually affected more with out of phase speakers, resulting in a bass light sound when in mono.

When we incorporate phase calculations in to sound we get a very complex picture of what is happening, as things may be different for every frequency component. With a simple case of two sources we can work out the resulting level using geometry. We know that if the waves are in phase, they will add, and if 180 degrees out of phase they will cancel each other.

By drawing two lines at the relevant angle to each other, and with their lengths proportional to their level, we can draw a third line connecting them and measure it to find the resulting level and phase angles relative to the originals.

Although this isn't particularly useful in the middle of a gig, it can be useful for working out speaker dispersion.

Interference

Interference occurs when two sound sources meet. For our purposes this is usually the effect of reflected sound on the direct sound, but is also applicable to multiple speaker sources. The main factor with interference is that it is not constant, but varies with the wavelength (hence frequency) and phase of the waves.

At a path difference of 17 cm (6.6 inches) there will be cancellation at 1 kHz (half wavelength), 3 kHz (1.5 times), 5 kHz (2.5 times) and so on. However at 2 kHz (whole wavelength) there is reinforcement, and also at 4 kHz (twice wavelength), 8 kHz (three times wavelength) and so on.

This violent series of cancellation and boosting is known as comb filtering, because on graph paper it would look a bit like the teeth of a comb. Comb filtering can reduce intelligibility and is also the principle behind flanging and phasing effects.

At other distances the effect is obviously different. With a seat fed from four speakers, the result is quite complex. Multiple speakers can cause this severe interference which is why ceiling speakers have to be positioned with care.

Microphones can suffer from similar effects from direct and reflected sound, such as from a lectern or table top, and this will affect the whole PA regardless of the speaker arrangement.

Propagation

Propagation is an important concept in PA as it is this which is responsible for reducing level as the audience gets further away. A point source (also called a monopole) radiates sound from a central point which expands in concentric spheres. As it does this it has to transfer its energy to an increasing area of air, which reduces its level.

If we take a cross section of the radiation at a point twice as far from the source as the original point, the area has increased by four times. At three times the distance, an area of nine times now has to be covered. So the sound energy is following the law that

energy = inverse square of the distance travelled

In other words, at double the distance it has reduced power by a quarter. However we are more interested in sound pressure rather than power. As sound pressure is equal to the square root of the power, maths tells us that the SPL will be directly proportional to the distance. So for double the distance it will have reduced by 6 dB.

The sound absorption due to travelling in air alone (i.e. if it is not allowed to disperse) is only 5 dB at 2 kHz for 150 m and 10 dB for 10 kHz at 30 m. Air absorption then isn't really our problem.

Hopefully, all this maths confirms that it is the loss of energy due to area *coverage* with distance, rather than the *distance* itself which causes the most loss. If we can stop the sound from dispersing and covering a greater area of air, then we can maintain more level. This is where the principle of the horn comes in. By preventing the coverage area from increasing we can make the sound travel further.

For a dipole (open backed) speaker, the SPL is a result of the sound coming from the front and the rear. This can be calculated using the following:

SPLr = SPL x cosine of listening angle

This shows us that with this type of speaker we lose about 10% at 25 degrees and 50% at 60 degrees. At 90 degrees (equal distance to front and rear) we theoretically get nothing. Level drop with distance is still 6 dB for a doubling of distance.

For a line source speaker, as used in the traditional column speaker, it radiates in concentric circles as the propagation is deliberately restricted in the vertical plane. This reduces the coverage area and we get only a 3 dB loss for a doubling in distance. Hence it can project twice as far for the same level.

Obstacles

What happens when sound meets an obstacle is another practical consideration. Sound can do a number of things, depending on its frequency and the size and absorption coefficient of the obstacle. Sound will either be reflected, diffracted, refracted, absorbed or transmitted.

Reflected

Sound behaves much like light when it meets a hard smooth reflective surface. When hitting a flat surface the angle of reflection will be the same as the angle of incidence. For a corner it will come out on a parallel path, regardless of the angle it entered. Sometimes a ghost image can be created which appears as far behind the surface as it is in front – this can cause comb filtering effects on its own.

If it hits a convex surface, the sound is reflected in all directions; if it hits a concave surface it is focused to some point, which is how parabolic microphones and satellite dishes work.

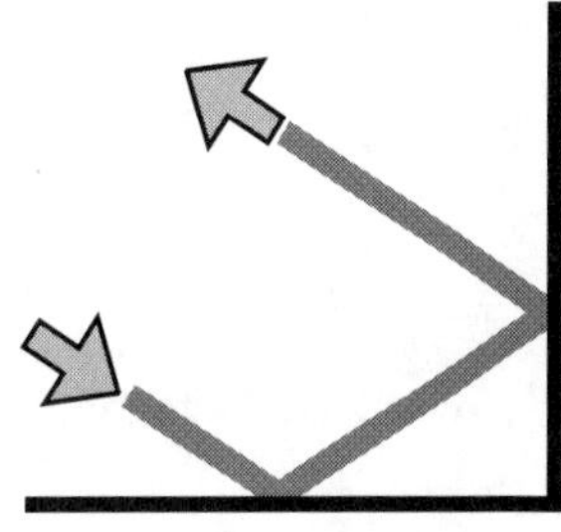

Top: Sound reflects off a hard surface
Bottom: Sound always reflects out of a corner at the same angle at which it arrived

Diffraction

Diffraction is when a sound flows around an obstacle (or meets a hole) that is much smaller than its wavelength. It basically ignores it and flows around it. If the wavelength of the sound is shorter than the obstacle then an acoustic shadow is created by the object.

Refraction

Refraction occurs when sounds travels between media of different densities, such as air and water, or air and solids, and even air at different temperatures. It bends towards the denser medium. For a venue, the hot air which rises to the ceiling tends to bend the sound back towards the audience.

Absorption

When sound hits certain materials (particularly treble in soft furnishing or curtains), the sound energy is absorbed and turned in to heat. The change in temperature is negligible so there is no danger of the furnishing catching fire at high sound volumes.

Transmission

Most transmission will occur when objects are in direct contact rather than air born. Providing some isolation between speakers and microphone stands and the floor can really help to tighten the bass response and reduce the chances of feedback and other interference.

Like most things with sound, the materials will exhibit different amounts of these effects at different frequencies, and that's when the fun starts.

The following table indicates the abilities of various materials to absorb sound energy. A figure of one is total absorption like an open field, while zero would be total reflection. You'll notice that it is different for different frequencies (usually more absorbent at higher frequencies).

Absorption coefficients of various materials

	125 Hz	250 Hz	500 Hz	1 kHz	2 kHz	4 kHz
Concrete	0.010	0.010	0.020	0.020	0.020	0.030
Brick	0.024	0.025	0.030	0.040	0.050	0.070
Plaster on brick	0.024	0.027	0.030	0.037	0.039	0.034
Plasterboard	0.300	0.300	0.100	0.100	0.400	0.400
Glass	0.030	0.030	0.030	0.030	0.020	0.020
Plywood (9 mm)	0.110	0.110	0.120	0.120	0.100	0.100
Plywood on 50 mm batten	0.350	0.250	0.200	0.150	0.050	0.050
Acoustic panels	0.150	0.300	0.750	0.850	0.750	0.400
Carpet (thick)	0.150	0.150	0.350	0.350	0.500	0.500
Curtains (heavy folds)	0.200	0.200	0.500	0.500	0.800	0.800
Fibreglass (50 mm)	0.190	0.510	0.790	0.920	0.820	0.780
Person (seated)	0.180	0.400	0.460	0.460	0.500	0.460

Reverberation

One of the many things that makes an event live is the sound of the environment it was performed in. For classical music, the reverb is an integral part of the music, and for any program material the absence of reverb sounds unnatural.

Unfortunately if the reverb is too long it can mush up the sound and reduce intelligibility. It is also part of the sound reservoir which can be responsible for feedback.

The audience will have a large damping effect on the reverb and needs to be accessed with them present. Reverb time is defined as the time it takes for a sound to decay to a thousandth of its original level (60 dB). Optimum reverb times are between 1.5 to 2.8 seconds for music and 0.5 to 1 second for speech.

Sound level SPL chart

120 dB	threshold of pain
70 dB	speech at 1 metre
0 dB	threshold of hearing

17

Connectors

Intro

At this point it is worth having a look at some of the connectors commonly used in live sound.

XLR 3 pin connectors

pin 1 = earth

pin 2 = hot (to IEC standard, pin 3 is hot on a Shure mic)

pin 3 = cold

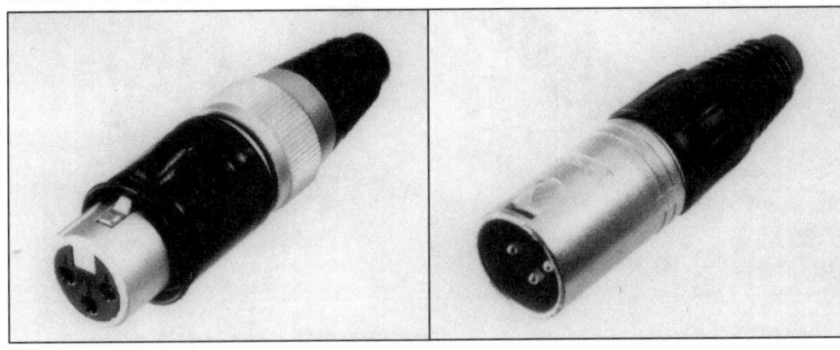

Mono jack 1/4 inch (6.35 mm)

Sleeve = earth

Tip = live

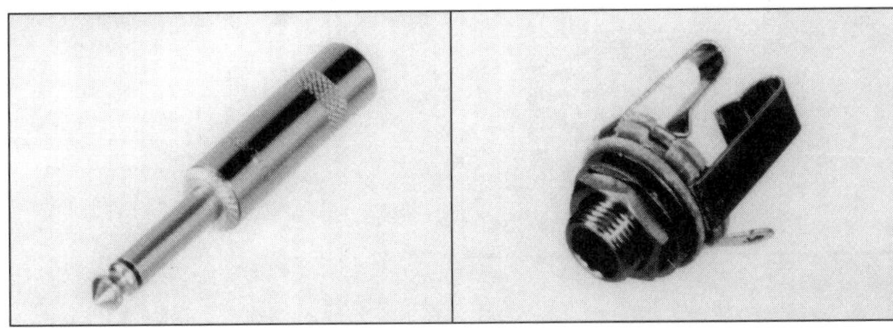

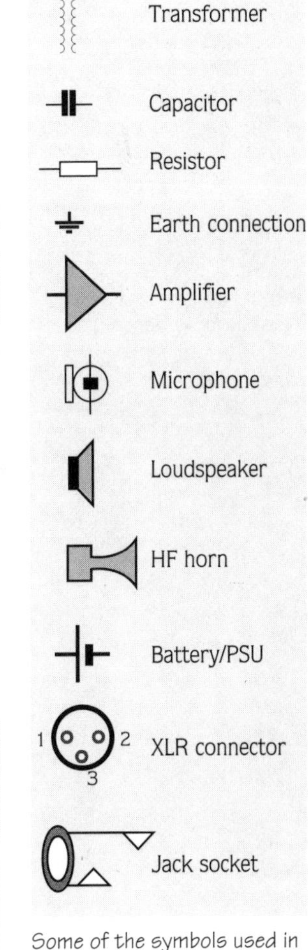

Some of the symbols used in this book

Left: female and male XLRs and mono jack plug and socket
(Pics courtesy Neutrik)

Stereo jack 1/4 inch (6.35 mm)

As stereo connector

sleeve = earth

ring = right channel (usually longer pin)

tip = left channel (usually shortest pin)

As insert lead

sleeve = earth

ring = send to device (can vary with some manufacturers)

tip = return from device

As balanced input/output

sleeve = earth

ring = cold

tip = hot

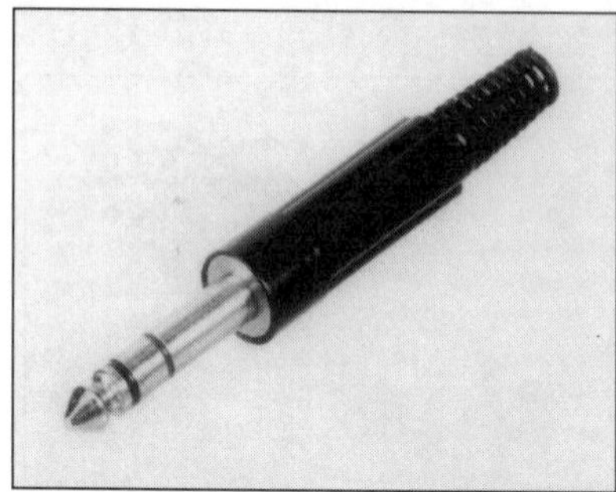

Stereo jack plug and socket (Pics courtesy Neutrik)

Phono plug connections

Sleeve = earth

Centre pin = signal

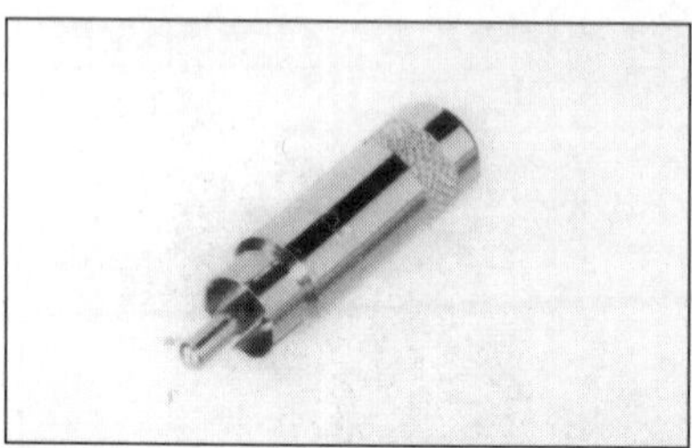

Phono plug (Pic courtesy Neutrik)

DIN plug 5 pin

pin no

1 left input

2 centre pin - earth

3 left output

4 right input

5 right output

(arranged in order of 1 4 2 5 3 clockwise, from solder side of plug)

Wiring standards for 5 pin DINS can vary widely between equipment.

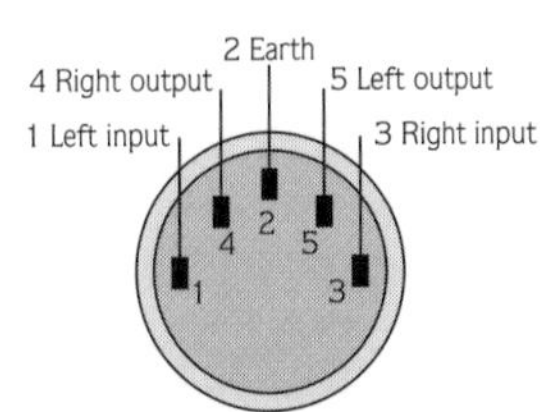

DIN plug from solder side (same as front view of chassis socket)

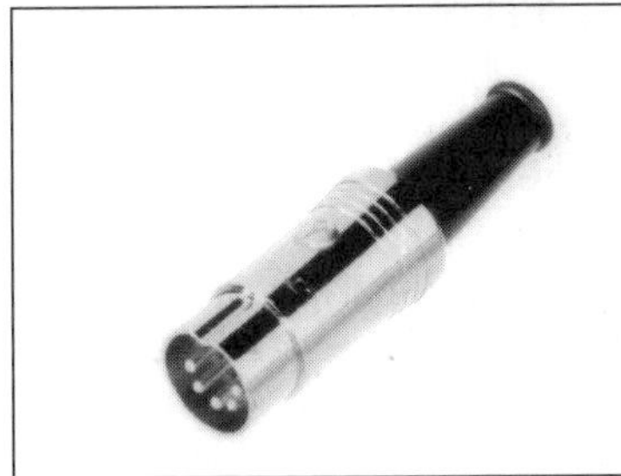

Left DIN plug and socket (Pics courtesy Neutrik)

Neutrik Speakon connections

This is available in 4 and 8 way configurations

	type	*1-*	*1+*	*2-*	*2+*	*3-*	*3+*	*4-*	*4+*
full range	nl4	ret	sig						
2 way	nl4	lf ret	lf sig	hf ret	hf sig				
	nl8	lf ret	lf sig	hf ret	hf sig				
3 way	nl8	lf ret	lf sig			mid ret	mid sig	hf ret	hf sig
4 way	nl8	lf ret	lf sig	lm ret	lm sig	hm ret	hm sig	hf ret	hf sig

(Key – sig = signal, ret = return, f = frequency, l = low, m = mid, h = high)

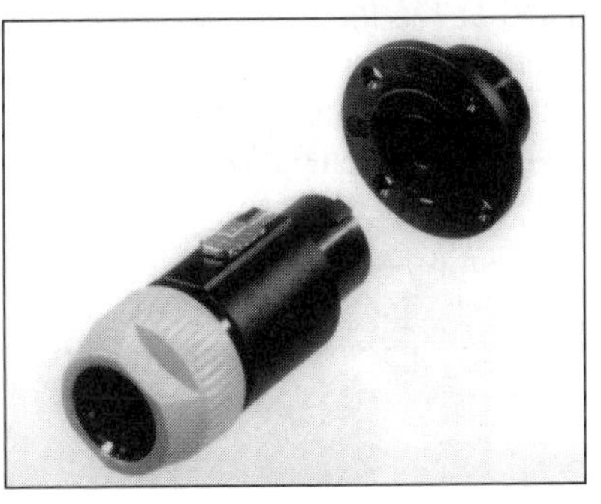

The Neutrik Speakon connector (Pic courtesy Neutrik)

18 Glossary

Intro

Every industry has its own unique set of 'techno' waffle. Here is your guide to talking live sound.

Active A device which requires some form of power. Usually offers superior performance to passive versions. See crossover.
ADT Artificial double tracking. A process which simulates the effect of performing the same thing twice. Introduces subtle variations which thicken the sound, making it stronger.
AFL After fade listen. Like solo but taken after the fader and pan pot allowing the effect of those controls to also be taken in to account.
Amplifier A device which increases the signal to higher levels. Power amplifiers provide sufficient volumes to drive loudspeakers.
Audience Members of the public who indirectly pay your wages containing several back seat (PA) drivers and drunks who want to spill beer over your gear, break or steal it. Handle with care and do not feed, these creatures are highly dangerous ;-).
Auxiliary A control on the mixer which takes a separate feed off a channel (and sometimes group) for use elsewhere. Importantly all similar auxiliaries are linked and share that destination. i.e. auxiliary 1 effects send to a reverb means all channels share that reverb. See pre fade, post fade and foldback.
Balanced line A system of interconnection involving two twisted cable cores and a screen. The cores are used as equal and opposite signal feeds and are much less prone to common mode interference such as hum and RF induced in the cable, as any noise is in phase in both cores and is cancelled at the input end by a transformer or electronic circuit. Requires a balanced connection at each end to function (i.e. microphone and balanced microphone input on mixer). Usually low impedance. The screen provides overall RF screening for the cable.
Band See *Performers.*
Bi-amplification A system where the signal from the mixer is split by an electronic crossover into two frequency ranges, each fed to a dedicated amplifier and speaker system. Provides a cleaner sound and reduces the chance of overdriving the speakers.

Cans Slang for headphones.
Capacitor mic A microphone which uses the capacitance between two plates (one fixed and charged, the other being the diaphragm) to produce electric current (alternative terms include condenser and electret).
Channel A vertical strip on a mixer. Input channels handle the input sources. and provide gain matching, in-built equalisation, routing (including to main outputs or groups, effects and foldback) and level control. Channels are repeated several times and once you understand one, you understand all those of that type.
Chorus An effect that recreates the effect of several people performing the same thing. Provides a thicker, richer, sweeter sound.
Crossover A device used to separate a signal into component frequency ranges so that the correct band is sent to the correct device (i.e. HF to tweeter). Passive crossovers use passive components (inductors and capacitors) to split the signal into its various components. An active crossover is placed between the mixer and the amplifiers.
Compression driver A special type of speaker that projects over a greater distance by limiting the spread of sound and focusing it.
Delay An effects device that creates a time lag between the original event and its version of it.
DI box Direct injection box. A device used to convert a line level signal source for use with a microphone type input. Converts the impedance and level to match and removes any phantom power. Also provides isolation, removing the chances of ground loops and hums. Active and passive versions are available.
Dispersion The range or area spread which a device can project to. In speakers the amount of the audience that can hear all the sound it produces.
Distortion Any variance from the input signal, usually associated with fuzzy or unclear sound reproduction. Caused by limited level response or poor linearity in electronics or mechanical systems like speakers.
Dongle A hardware accessory supplied with a software program to limit its unauthorised use. The software will not run without this device.
Driver Slang for speaker.
Dry hire Equipment hire without an operator included.
Echo A device which repeats the sound several times with decreasing volume. Similar to shouting in a canyon.
EQ Equalisation. The process of changing the frequency balance of a signal, by increasing or decreasing the level of parts of it. Akin to the tone controls on a hifi but usually offering more control.
Fader The sliding device on a mixer used to control the level.
Feedback The howling sound which occurs when the signal from a speaker is fed back into a microphone feeding it, causing extreme excitation in the system due to cyclic amplification.
Filter A type of equalising circuit that removes certain frequencies. See *LPF*, *HPF* and *Notch filter*.
Flanging An effect that reproduces the sound of comb filtering – like listening to music at the end of a long tube.
FOH Front of house. The area concerned with the audience.

Foldback The signal mix sent to the performers on-stage. Normally fed to stage monitors or recently to in-ear foldback systems.
Fuse A device used to protect electronic equipment from fire due to excessive current drain. A fuse *must* only be replaced with the same value or severe risks are being taken. A fuse can blow because of age or shock, or because of a fault in the system. You'll soon find out which when you replace it.
Frequency A term used to describe the rate of regular cyclic change in a sound. Perceived by the ear as the pitch.
Graphic equaliser A device that consists of a tone control for several bands. Each band can be adjusted simultaneously providing a very wide scope for frequency response change. The level of each frequency band is controlled by a sliding fader which visually (graphically) indicates what the equalisation setting is. Used for extreme tone changes or in PA systems on the main outputs to correct for the frequency response of the venue and PA system, and to help 'tune out' feedback.
Group A fader control on the mixer which affects the final level of any channels routed to it. It controls a group of channels, hence its name.
Headphones Small speakers worn close to the ears for personal monitoring.
HF High frequencies – as in treble. The right hand range of notes on a piano.
HPF High pass filter. A device which attenuates the low frequency range but allows the higher frequency range to pass unaltered.
Key disk Another means of software protection (see dongle) that limits the use of the software to the user holding the key disk. The key disk has to be inserted every time the program is used, although some install a non transferable copy onto the hard drive.
LPF Low pass filter. Allows low frequencies to pass unaffected and attenuates high frequencies.
Microphone A transducer used to convert acoustic energy (air vibrations) into a signal the mixer will understand (electricity).
Mixer A device used to control and route nearly all of the other PA equipment. Comprises lots of vertical strips called input channels (which are just repeated (when you understand one you understand them all) and an output section. Provides level and impedance matching, built-in equalisation, the facility to add external effects, and route the signal to various destinations, including foldback monitors, recording feeds and main FOH amplifiers.
Noise Any signal which is not required. In a PA system the hiss and hum which is not part of the music.
Notch filter A device which removes a very narrow range of frequencies. Often used in feedback rejection or to eliminate spot frequencies such as tape hiss or mains hum.
Parametric A type of equalisation circuit which provides swept frequency and Q (bandwidth) parameters in addition to level boost and cut.
Passive A device requiring no electrical power. See *Crossover*.
Peak The maximum part of a signal. The loudest it ever gets before distortion.

Performers The people who indirectly pay your wages and book you again (or not!).
PFL Pre fade listen. Another word for solo. See *Solo.*
Phantom power A DC voltage fed down the microphone cable without using any extra cable conductors. Provides 50 V DC for capacitor microphones and active DI boxes (and some talkback systems).
Phase On a waveform, the point through the cycle relative to another. Speakers and microphones are said to be in-phase when they both push or pull when the same signal is applied to them.
Phasing See *phase.*
Pink noise A special type of noise used to test an audio system. Less bright than white noise, it consists of random frequencies of equal energy in each octave (i.e. the same energy between 200 – 400 Hz as 1000 – 20,000Hz)
Post fade On the mixer, a feed of the signal taken after the fader. It is always proportional to the fader position and decreases and increases in level with the fader. Usually used for effects sends so that the effect retains its balance with the dry signal.
Pre fade On the mixer, a feed of the signal taken before the fader. Allows independent use ignoring the position of the channel fader. Normally used for foldback.
Reverb A device to simulate the acoustics and ambience of an environment i.e. hall or room.
RMS Root mean square. A measurement of the average level from a device. A more realistic measurement than peak when used to rate amplifier and speaker power.
Slew rate The ability of an electronic device to respond quickly to changes in the input signal level. Particularly shows up in the HF range.
Solo in place A special mode on some mixers which deliberately feeds the signal during solo to the main outputs. Useful as an inverse muting button. Dangerous in PA systems and is more suited to recording mixers where inappropriate use can just be remixed.
Sub group Part of the mixer used to control the final level of a group of channels.
Sub grouping Formed by using the group fader to feed back in to the stereo mix,
Solo Facility on a mixer to listen to individual signals without affecting the main signal path. Does not affect the main output feeds
Sweepable A type of equalisation where a frequency tuning control is provided for the user, in addition to the level boost/cut control. Provides far more flexibility in getting the sound you want, otherwise you are at the mercy of the manufacturer's designer. See *Parametric.*
Thermal Related to heat. Thermal overload of an amplifier means it has run too hot and safety devices have switched in to allow the amp to cool down, hopefully before developing a permanent fault.
Time delay When speakers are some physical distance apart, a time delay is used to realign the sound from them compared to the original source. Amounts to 1 ms per 30 cm (foot) of distance.

Transducer Any device which converts one form of energy into another; e.g. microphones and loudspeakers.

Transient The wavefront of a signal responsible for the attack of a sound. Mainly affected by poor high frequency response and low slew rates. A major factor in the equality, cleanness and transparency of a sound.

Tri-amplification A system where the signal from the mixer is split by an electronic crossover in to three separate frequency ranges, each of which feeds a dedicated amplifier and loudspeaker system. Produces a cleaner sound with less risk of speakers blowing.

Tweeter A loudspeaker specifically designed to reproduce the treble range of the audio spectrum.

Wavelength The distance between two identical points of a cycle of a waveform; e.g. peak to peak.

White noise A special type of noise signal containing equal energy at all frequencies in each octave. Brighter than pink noise. Similar to radio station tuning. Used in synthesis for wind and breath noises.

Woofer A loudspeaker specifically designed to handle the low frequency range of the audio spectrum.

19

Contacts

Professional bodies

AES
PO Box 645
Slough
Berks SL1 8BJ
Tel 01628 663 725
Fax 01626 667 002

APRS
2 Windsor Square
Silver Street
Reading
Berks RG1 2TH
Tel 01734 756 218
Fax 01734 756 216

ASCAP
Suite 10/11
52 Haymarket
London SW1Y 4RP
Tel 0171 973 0069
Fax 0171 973 0068

MCPS
Elgar House
41 Streatham High Road
London SW16 1ER
Tel 0181 769 4400
Fax 0181 769 8792

Musicians Union
60-62 Clapham Road
London SW9 0JJ
Tel 0171 582 5566
Fax 0171 582 9805

PRS
29/33 Berners Street
London W1P 4AA
Tel 0171 580 5544
Fax 0171 631 4138

Sound & Comms Industries Federation
4b High Street
Burnham, Slough
Berks SL1 7JH
Tel 01628 667 633
Fax 01628 665 882

Vpl Ltd
Ganton House
14/22 Ganton Street
London W1V 1LB
Tel 0171 437 0311
Fax 0171 734 9797

Press and publishing

Applause Publications
132 Liverpool Road
London N1 1LA
Tel 0171 700 0248
Fax 0171 700 0301

Audio Media
Media House
Burrell Road
St Ives
Cambs PE17 4LE
Tel 01480 461244
Fax 01480 492422

Lighting and Sound International
7 Highlight House
St Leonards Road
Eastbourne
East Sussex BN21 3UH
Tel 01323 642 639
Fax 01323 646 905

Live!
35 High Street
Sandridge
St Albans
Herts AL4 9DD
Tel 01727 843 995
Fax 01727 844 417

PC Publishing
Export House
130 Vale Road
Tonbridge TN9 1SP
Tel 01732 770893
Fax 01732 770268

Pro Sound News
Ludgate House
245 Blackfriars Road
London SE1 9UR
Tel 0171 620 3636
Fax 0171 921 5984

Sound and Communication Systems
Pirate Publishing
3/9 Broomhill Road
London SW18 4JQ
Tel 0181 871 5258
Fax 0181 877 1940

Sound on Stage
Media House
Trafalgar Way, Bar Hill
Cambridge
Cambs CB3 8SQ
Tel 01954 789888
Fax 01954 789895

Manufacturers

Audio Projects
Unit 8, Speedgate Farm
Mussendal Lane
Fawkham
Kent DA3 8NJ
Tel 01474 879446
Fax 01474 872925

Audio Technica Ltd
Unit 2 Royal London Trad. Est.
Old Lane
Leeds
Yorks LS11 8AG
Tel 0113 277 1441
Fax 0113 270 4836

Autograph Sales Ltd
102 Grafton Road
London NW5 4BA
Tel 0171 485 3749
Fax 0171 485 0681

Beyer Dynamic Ltd
Unit 14
Cliffe Ind. Est.
Lewes Sussex BN8 6JL
Tel 01273 479 411
Fax 01273 471 825

Beyma Ltd
Unit 10 Acton Vale Ind. Park
Crowley Road
London W3 7QE
Tel 0181 749 7887
Fax 0181 749 9875

Bruel & Kjaer Ltd
92 Uxbridge Road
Harrow
Middx HA3 6BZ
Tel 0181 954 2366
Fax 0181 954 9504

Bss Audio Ltd
Linkside House
Summit Road
Cranbourne Ind. Est.
Potters Bar
Herts EN6 3JB
Tel 01707 660 667
Fax 01707 660 755

Carlsbro Electronics
Cross Drive
Kirkby-in-ashfield
Notts NG17 7LD
Tel 01623 753 902
Fax 01623 755 436

Clair Bros. Sensible Music Ltd
Unit 10
105 Blundell Street
London N7 9BN
Tel 0171 700 6655
Fax 0171 609 9478

EAW - Eastern Acoustic Works
721 Tudor Estate
Abbey Road
Park Royal
London NW10 7UN
Tel 0181 961 6858
Fax 0181 961 6857

EMO Systems Ltd
Durham Road
Ushaw Moor
Durham City
Co. Durham DH7 7LF
Tel 0191 373 0787
Fax 0191 373 3507

ESS
Unit 14 Bleakhill Way
Hermitage Land Ind. Est.
Mansfield
Notts NG18 5EZ
Tel 01623 647 291
Fax 01623 22500

Garwood Communications
136 Cricklewood Lane
London NW2 2DP
Tel 0181 452 4635
Fax 0181 452 6974

Harman Audio (AKG)
Unit 2 Borehamwood Ind. Park
Rowley Lane
Borehamwood
Herts WD6 5PZ
Tel 0181 207 5050
Fax 0181 207 4572

HW International (Shure)
167-171 Willoughby Lane
Brantwood Ind. Area
London N17 0SB
Tel 0181 808 2222
Fax 0181 808 5599

John Henry Enterprises
16-24 Brewery Road
London N7 9NH
Tel 0171 609 9181
Fax 0171 700 7040

Klark Teknik Plc
Klark Ind. Park
Walter Naxh Road
Kidderminster Worcs DY11 7HJ
Tel 01562 741 515
Fax 01562 745 371

Klotz/vdc Trading
Units 1 & 2
43 Carol Street
London NW1 0HT
Tel 0171 284 1444
Fax 0171 482 4219

LMC Audio Systems Ltd
Unit 10 Acton Vale Ind. Park
Cowley Road
London W3 7QE
Tel 0181 743 4680
Fax 0181 749 9875

Martin Audio
19 Lincoln Road
High Wycombe
Bucks HP12 3RD
Tel 01494 535 312
Fax 01494 438 669

Music Lab Ltd
72-74 Eversholt Street
London NW1 1BY
Tel 0171 388 5392
Fax 0171 388 1953

Neutrik
Columbia Business Park
Sherbourne Ave
Ryde
IOW P033 3QD
Tel 01983 811441
Fax 01983 811 439

P&R Audio
4 Swan Business Centre
Station Road
Hailsham
Sussex BN27 2BY
Tel 01323 849522
Fax 01323 849533

Peavey Electronics
Hatton House
Hunters Road
Corby
Northants NN17 5JE
Tel 01536 205 520
Fax 01536 269 029

Samson Technologies
154 Clapham Park Road
London SW4 7DE
Tel 0171 498 4861
Fax 0171 498 4861

Sennheiser Uk Ltd
12 Davies Way
Knaves Bleech Bus. Centre
Loudwater
High Wycombe HP10 9QY
Tel 01628 850 811
Fax 01628 850 958

Shuttlesound Ltd
4 The Willows Centre
Willow Lane
Mitcham
Surrey CR4 4NX
Tel 0181 646 7114
Fax 0181 640 7583

Soundcraft Electronics Ltd
Cranbourne House / Ind. Est.
Cranbourne Road
Potters Bar
Herts EN6 3JN
Tel 01707 665 000
Fax 01707 660 482

Soundtracs Plc
91 Ewell Road
Surbiton
Surrey KT6 6AH
Tel 0181 399 3392
Fax 0181 399 6821

Studiomaster Diamond Ltd
Studiomaster House
Chaul End Lane
Luton
Beds LU4 8EZ
Tel 01582 570 370
Fax 01582 494 343

Toa Corporation Plc
Tallon Road
Hutton Ind. Est.
Hutton
Essex CM13 1TG
Tel 01277 233 882
Fax 01277 233 566

Trace Elliot Ltd
The Causeway
Maldon
Essex CM9 7GG
Tel 01621 851 851
Fax 01621 851 932

Trantec/bbm Ltd
30 Wates Way
Mitcham
Surrey CR4 4HR
Tel 0181 640 1225
Fax 0181 640 4896

Turbosound Ltd
Star Road
Partridge Green
Sussex RH13 8RY
Tel 01403 711 447
Fax 01403 710 155

Washburn Uk Ltd
Arnor Way
Letchworth
Herts SG6 1UG
Tel 01462 482 466
Fax 01462 482 997

Wembley Loudspeakers
Unit A4
Askew Crescent Workshops
London W12 9DP
Tel 0181 743 4567
Fax 0181 749 7957

Yamaha Kemble Ltd
Sherbourne Drive
Tilbrook
Milton Keynes
Bucks MK7 8BL
Tel 01908 366 700
Fax 01908 368 872

Venues

100 Club
100 Oxford Street
London W1N 9FB
Capacity 300
Tel 0171 0933

Albany Empire
Douglas Way
Deptford
London SE8 4AG
Capacity 425
Tel 0181 692 4446

Alexandra Palace
Alexandra Palace Way
Wood Green
London N22 4AY
Capacity 7250
Tel 0181 365 2121
Fax 0181 883 3999

Barbican Centre
Barbican
London EC2Y 8DS
Capacity 1995
Tel 0171 638 4141
Fax 0171 638 7832

Brixton Academy
211 Stockwell Road
Brixton
London SW9 9SL
Capacity 4272
Tel 0171 274 1525
Fax 0171 738 4427

Break For The Border
5 Goslett Yard
125-127 Charing Cross Road
London WC2H 0CA
Capacity 300
Tel 0171 437 8595
Fax 0171 437 0479

Camden Palace
1 Camden High Street
London NW1 7JE
Capacity 1500
Tel 0171 387 0428
Fax 0171 388 8850

Dingwalls
Jongleurs Middle Yard
Off Chalk Farm Road
Camden Town
London NW1 8AB
Capacity 500
Tel 0171 267 1577
Fax 0171 267 0586

The Fridge
Town Hall Parade
Brixton Hill
London SW2 1RJ
Capacity 1800
Tel 0171 326 5100
Fax 0171 274 2879

Half Moon
93 Lower Richmond Road
Putney
London SW15 1EU
Capacity 150
Tel 0181 780 9383
Fax 0181 789 7863

Hippodrome
Hippodrome Corner
Leicester Square
London WC2 7JH
Capacity 1650
Tel 0171 437 4837
Fax 0171 434 4225

Jazz Cafe
5 Parkway
London NW1 7PG
Capacity 350
Tel 0171 916 6060
Fax 0171 916 6622

Jongleurs
49 Lavender Gardens
Clapham
London SW1 1DJ
Capacity 350
Tel 0171 924 2248
Fax 0171 924 5175

Kings Head
4 Fulham High Street
London SW6 3LQ
Capacity 200
Tel 0171 1413

London Palladium
Argyll Street
London W1
Capacity 2298
Tel 0171 437 6678

Marquee Club
105 Charing Cross Road
London WC2H 0DT
Capacity 850
Tel 0171 437 6603
Fax 0171 434 1651

Mean Fiddler
22-28a High Street
Harlesden
London NW10 4LX
Capacity 600 + 150
Tel 0181 961 5490
Fax 0181 961 9238

Royal Albert Hall
Kensington Gore
London SW7 2AP
Capacity 5200
Tel 0171 589 3203
Fax 0171 823 7725

Royal Festival Hall
Belvedere Road
London SE1 8XX
Capacity 2900
Tel 0171 921 0600
Fax 0171 401 8834

Ronnie Scotts Club
47 Frith Street
London W1D 5HT
Capacity 300
Tel 0171 439 0747
Fax 0171 437 5081

The Roadhouse Jubilee Hall
35 The Piazza
Covent Garden
London WC2
Capacity 400
Tel 0171 240 4016

Wembley Conference Centre
Wembley Stadium
Middx HA9 0DW
Capacity 2700
Tel 0181 902 8833
Fax 0181 903 3234

Wembley Stadium
Wembley
Middx HA9 0DW
Capacity 72000
Tel 0181 902 8833
Fax 0181 900 1055

Wigmore Hall
36 Wigmore Street
London WC1H 9DF
Capacity 540
Tel 0171 486 1907
Fax 0171 224 3800

Rigging

All Events UK Ltd
Moelwyn House
301 Percy Street
Billericay
Essex CM12 0RB
Tel 01277 651 335
Fax 01277 650 670

Cochrane Promotions Ltd
18 St Albans Road
Dartford
Kent DA1 1TF
Tel 01322 229 923
Fax 01322 284 145

Garwood Communications
136 Cricklewood Lane
London NW2 2DP
Tel 0181 452 4635
Fax 0181 452 6974

Outback Rigging Ltd
11 Kendall Court
Kendal Avenue
Park Royal
London W3
Tel 0181 993 0066
Fax 0181 752 1753

Unusual Rigging Ltd
4 Dalston Gardens
Stanmore
Middx HA7 1DA
Tel 0181 206 2733
Fax 0181 206 1432

PA hire

Audio and Acoustics Ltd
United House
London Road
London N7 9DP
Tel 0171 700 2900
Fax 0171 700 6900

Audiolease Ltd
Unit 5 Lion Works
Station Road East
Whittlesford
Cambs CB2 4NL
Tel 01223 837 775
Fax 01223 834 848

Better Sound Ltd
33 Endell Street
London WC2H 9BA
Tel 0171 836 0033
Fax 0171 497 9285

Britannia Row Productions Ltd
9 Osiers Road
London SW18 1NL
Tel 0181 877 3949
Fax 0181 874 0182

Concert Systems
Unit 4d Stag Ind. Est.
Atlantic Street
Altrincham Cheshire WA14 5DW
Tel 0161 927 7700
Fax 0161 927 7722

Dimension Audio Ltd
Unit 3
307 Merton Road
London SW18 5JS
Tel 0181 877 3414
Fax 0181 877 3410

Efx Audio
Unit 5 Lutton Court
Lutton Place
Edinburgh
Scotland EH8 9PD
Tel 0131 667 6127
Fax 0131 667 6127

Entec Sound & Light
517 Yeading Lane
Northolt
Middx UB6 6LN
Tel 0181 842 4004
Fax 0181 842 3310

ESE Audio
Briar Knoll Chapel Lane
Sissinghurst
Kent TN17 2JN
Tel 01580 715 039
Fax 01580 713 419

ESS PA Hire
Unit 14 Bleakhill Way
Hermitage Lane Ind. Est.
Mansfield
Notts NG18 5EZ
Tel 01623 647 291
Fax 01623 22500

Hearhouse Ltd
17 Penn Street
Birmingham
West Midlands B4 7RJ
Tel 0121 333 3390
Fax 0121 333 3347

JHE Audio Ltd
16-24 Brewery Road
London N7 9NH
Tel 0171 609 9181
Fax 0171 700 7040

Sensible Music Ltd
Unit 10
105 Blundell Street
London N7 9BN
Tel 0171 700 6655
Fax 0171 609 9478

SSE Hire Ltd
201 Coventry Road
Birmingham
West Midlands B10 0RA
Tel 0121 766 7170
Fax 0121 766 8217

Strawberry Rental Services
3 Waterloo Road
Stockport
Cheshire SK1 3BD
Tel 0161 477 6270

Theatre Projects Sound Sevices Ltd
13 Field Way
Greenford
Middx UB6 8UN
Tel 0181 566 644
Fax 0181 566 6365

Wigwam Acoustics Ltd
St Annes House
Ryecroft Avenue
Heywood
Lancs OL10 1QB
Tel 01706 624 547
Fax 01706 365 565

Mobile toilets

Mobiloo Ltd
Russell Street
Kettering
Northants NN16 0EL
Tel 01536 410 863
Fax 01536 519 758

Poly-loo Enterprises Ltd
Unit 2 Princes Road
Montagu Ind. Est.
Montagu Road
London 3PR
Tel 0181 884 2838
Fax 0181 807 0280

Superloo Ltd
Unit 5 Smallwood Lane
Smallwood
St Albans
Herts AL4 0LL
Tel 01727 822 120
Fax 01727 822 886

Lighting hire

Alien Products Ltd
80 Bousley Rise
Ottershaw
Chertsey
Surrey KT16 0LB
Tel 01932 872 454
Fax 01932 872 909

Black Light Ltd
18 West Harbour Road
Granton
Edinburgh
Scotland EH5 1PN
Tel 0131 551 2337
Fax 0131 552 0370

Concert Lights Uk Ltd
Undershaw Works
Brookside Road
Bolton
Lancs BL2 2SF
Tel 01204 391 343
Fax 01204 363 238

Donmar Ltd
54 Cavell Street
Whitechapel
London E1 2HP
Tel 0171 790 9937
Fax 0171 790 6634

Entec Sound And Light
517 Yeading Lane
Northolt
Middx UB5 6LN
Tel 0181 842 4004
Fax 0181 842 3310

Eurosound Ltd
Unit 1 Intake Lane
Woolley
Wakefield
Yorks WF4 2LG
Tel 01484 866 066
Fax 01484 866 299

Lightworks Ltd
2a Greenwood Road
London E8 1AB
Tel 0171 249 3627
Fax 0171 254 0306

Lite Alternative Ltd
Unit 402
Phoenix Park Ind. Est.
Heywood
Lancs
OL10 2JG
Tel 01706 627 066
Fax 01706 627 068

Generators

Backroom Rentals
73 Noyna Road
Tooting Bec
London SW17 7PQ
Tel 0181 767 5404
Fax 0181 672 2870

Northern Light
35-41 Assembley Street
Leith
Edinburgh
Scotland EH6 7RG
Tel 0131 553 2383
Fax 0131 553 3296

Sensible Music Ltd
Unit 10
105 Blundell Street
London N7 9BN
Tel 0171 700 6655
Fax 0171 609 9478

Flightcases

5 Star Cases
10-12 Sandall Road
Wisbech
Cambs PE13 2RS
Tel 01945 474 080
Fax 01945 64416

Cp Cases
Unit 11 Worton Hall Ind. Est.
Worton Road
Isleworth
Middx TW7 6ER
Tel 0181 568 1881
Fax 0181 568 1141

Cripple Creek Case Co. Ltd
Devonshire Works
Barley Mow Passage
Chiswick
London W4 4PH
Tel 0181 995 5348
Fax 0181 747 8704

Midland Custom Cases
19 Jameson Road
Aston
Birmingham B6 7SK
Tel 0121 326 6456
Fax 0121 327 6928

Pyrotechnics

Artem Visual Effects
Perivale Industrial Park
Clausen Way
Perivale
Middx UB6 7RH
Tel 0181 997 7771
Fax 0181 997 1503

Laser Hire
30 Water Street
Birmingham
West Midlands B3 1HL
Tel 0121 236 2243
Fax 0121 236 0764

Le Maitre Fireworks Ltd
Fourth Drove
Fengate
Peterborough
Cambs PE1 5UR
Tel 01733 346 824
Fax 01733 68619

Theatrical Pyrotechnics
The Loop, Manston Airport
Ramsgate
Kent CT12 5DE
Tel 01843 823 545
Fax 01843 822 655

Vulcan Fireworks Uk Ltd
52 North Street
Carlshalton
Surrey SM5 2HH
Tel 0181 669 4178
Fax 0181 773 0305

Hire

Audiohire
2 Langler Road
Kensal Rise
London NW10 5TL
Tel 0181 960 4466
Fax 0181 458 6148

Advanced Sounds Ltd
London
Tel 0181 462 6261
Fax 0181 462 8621

Ess Pa Hire
Unit 14 Bleakhill Way
Hermitage Lane Ind. Est.
Mansfield Notts NG18 5EZ
Tel 01623 647 291
Fax 01623 22500

Euro Hire
Lynchford Lane
Lynchford Road North Camp
Farnborough
Hants GU14 6JD
Tel 01252 511 407
Fax 01252 373 531

Fx Rentals Ltd
Unit 3 Park Mews
213 Kilburn Lane
London W10 4BQ
Tel 0181 964 2288
Fax 0181 964 1910

John Henry Enterprises
16-24 Brewery Road
London N7 9NH
Tel 0171 609 9181
Fax 0171 700 7040

Maurice Placquet
110-112 Disraeli Road
London SW15 2DX
Tel 0181 870 1335
Fax 0181 877 1036

Music Lab Hire
72-74 Eversholt Street
London NW1 1BY
Tel 0171 388 5392
Fax 0171 388 1953

Nomis Studios
45-53 Sinclair Road
London W14 0NS
Tel 0171 602 6351
Fax 0171 603 5941

Northern Backline Hire
Unit 16 Brighton Road Ind. Est.
Heaton Norris
Stockport
Cheshire SK4 2BE
Tel 0161 431 4127
Fax 0161 431 4127

Phantom Power P A & Productions
1 Yorke Street
Burnley
Lancs BB11 1HD
Tel 01282 839 800
Fax 01282 839 779

Strawberry Rental Services
3 Waterloo Road
Stockport
Cheshire SK1 3BD
Tel 0161 477 6270
Fax 0161 476 0060

Studiohire
8 Daleham Mews
London NW3 5DB
Tel 0171 431 0212
Fax 0171 431 1134

Swan Music Ltd
Unit 3 Plymouth Crt. Bus. Centre
166 Plymouth Grove
Manchester M13 0AF
Tel 0161 273 3232
Fax 0161 274 4111

Videowalls

Electrosonic
Hawley Mill Hawley Road
Dartford
Kent DA2 7SY
Tel 01322 222 211
Fax 01322 282 282

Laserpoint Comms Ltd
44 Clifton Road
Cambridge
Cambs CB1 4FD
Tel 01223 212 331
Fax 01223 214 085

Metro Video
The Old Bacon Factory
57-59 Great Suffolk Street
London SE1 0BS
Tel 0171 928 2088
Fax 0171 261 0685

Proquip Rental Ltd
Unit B The Forum
Hanworth Lane
Chertsey
Surrey KT16 9JX
Tel 01932 567 111
Fax 01932 569 333

REW Communication Services
Unit 4 Leatherhead Ind. Est.
Station Road
Leatherhead
Surrey KT22 7AG
Tel 01372 361 291
Fax 01372 361 293

Index